UNITED STATES COAST GUARD CUTTER SHERMAN (WHEC-720) CIRCUMNAVIGATION DEPLOYMENT 2001

13 JANUARY – 13 JULY

BY

Lieutenant Edward Leo Semler Jr., USCG (Retired)

Copyright © 2016 by Edward Leo Semler Jr.

All rights reserved by the author. This publication is a historical record for all, and any part may be reproduced for non-monetary educational purposes.

First Edition: 2016

Revised in: 2020

Revised in 2022 to include *Sherman's* Ship Logs

Library of Congress Control Number: 2016901225

ISBN: 978-0-692-62366-4

Printed in the United States of America

Cover layout by Edward L. Semler Jr.

To the crew of the *Sherman* and their families.

TABLE OF CONTENTS

Introduction	1
Preparation for Deployment	3
Enroute to the Persian Gulf	13
The Persian Gulf	27
Heading Home to Alameda, California	61
USCGC Sherman Circumnavigation Crew	99
USCGC Sherman Ships Logs	107
Sources and References	255
About the Author	259

INTRODUCTION

While writing "Around the World," my memoir of active duty service, I had touched briefly on *Sherman's* 2001 deployment to the Persian Gulf and circumnavigation of the globe. The circumnavigation is such an interesting and unusual event that I wanted to document it further and establish a permanent written record for all to read and enjoy. During this deployment my primary job was the engineering division main propulsion chief. I also held the collateral duties of command enlisted advisor and health and comfort boarding team member. So I had a good perspective of the events leading up to the deployment and subsequent circumnavigation.

Even though I personally made the deployment and had pictures, personal journals, and video, it would have been impossible for me to have compiled this book without help. The U.S. Coast Guard Office of Cutter Forces and the U.S. Navy provided me with Freedom of Information Act (FOIA) documents. Those documents included *Sherman's* ship's logs, *Sherman's* operation summary, and Navy command history reports. Whenever I came upon a discrepancy I always fell back on the ship's logs, which are included in the back of this book, as being the legitimate documentation. And perhaps my most important source of information came by way of my fellow shipmates who provided me with interviews and valuable data they had saved over the years.

One final bit of morning cleanups before we shove off, I mostly refer to the Persian Gulf by that name. But it is also referred to in this book as the Arabian Gulf -- the names are interchangeable. The military likes to refer to it as the Arabian Gulf so as not to offend Arab countries in the region because the name Persian Gulf, to them, infers ownership by the country of Iran.

With all of the data I had collected I contemplated how I should approach this book. As a historical record it needed to detail events. But this can make for a dry and boring read. I felt that by adding personal accounts and commentary it would give it life. My ultimate goal is to provide both a fond memory to those who made the deployment and a historical record of the event.

PREPARATION FOR DEPLOYMENT

In 2001 the responsibility of enforcing United Nations sanctions against the country of Iraq in the Persian Gulf was typically a U.S. Navy function, with the help of other U.N. members. The Navy had also been utilizing Coast Guard high endurance cutters to assist them for several years. This was possible under a memorandum of understating, which stated Coast Guard assets could be utilized for certain Navy missions to include maritime interception operations and peacetime military engagements. This was nothing new. Coast Guard assets had been used widely in World War II and in Vietnam. And during times of war, the Coast Guard falls under the direction of the Navy.

Therefore, when the Navy needed to replace the *USS Ingraham (FFG-61)* for an upcoming six-month deployment they looked to the Coast Guard to supply a replacement asset. *Sherman* was notified by Coast Guard Headquarters that she would be replacing *Ingraham* and deploying January 2001.

She would be the fourth high endurance cutter to deploy to the Persian Gulf enforcing U.N. sanctions on Iraq. The *USCGC Morgenthau (WHEC-722)* was the first in 1996, the *USCGC Chase (WHEC-718)* was the second in 1998, and the *USCGC Midgett (WHEC-726)* was the third in 2000.

Home ported at the Coast Guard Island in Alameda, California she was an older cutter and commissioned in 1968. She was the sixth cutter in the "Secretary" class of cutters and named in honor of John Sherman, the 32nd Secretary of the Treasury serving from 1877 to 1881. He would later go on to become the 35th Secretary of State from 1897 to 1898 under President McKinley. The "Secretary" class was simply referred to as the "378" class within the Coast Guard which was a reference to the cutters length.

She was no stranger to operating in hostile areas and had completed a 1-year tour in Vietnam in 1970 to 1971. During that deployment *Sherman* was assigned to Coast Guard Squadron Three. Her tasking during the war was primarily in support of Operation Market Time, which involved sorting through hundreds of small vessels off the Vietnamese coast in search of enemy weapons smugglers. *Sherman's* crew inspected some 900 vessels during her 10-month tour in Southeast Asia. The older 5" gun – now replaced by a 76mm mount, answered 152 calls for naval gunfire support, including a running fight on the night of 21 November 1970 which resulted in the sinking the North Vietnamese armed freighter *SL-3,* which was carrying tons of enemy munitions. This is considered to be the last enemy warship to be sunk in combat by a U.S. vessel.

Compared to her Navy counterparts *Sherman* was minimally armed, which is understandable because she was not designed as a warship. She did however have a 76mm cannon mounted on the bow, two 25mm guns (one mounted on either side amidships), six .50 caliber machine guns mounted around the vessel, and a 20mm close-in weapons system (CIWS) mounted on the fantail.

But the Navy didn't want the Coast Guard in the Persian Gulf because of anything the 378's brought to the table in the way of fire power. They wanted the Coast Guard there because of their superior law enforcement boarding skills and *Sherman's* relatively shallow draft. *Sherman* would be tasked to operate in the Northern Arabian Gulf (NAG) which has areas prohibitive to the other Middle Eastern Forces (MEF) vessels. The water in some areas of operation, near the Iranian Territorial Seas, have shoals with depths between 20 and 28 feet. *Sherman* would be able to navigate these areas pursuing suspected smugglers with good self-defense and sprint speed with her turbine propulsion.

As for personnel, *Sherman* had a mixed gender crew which was common with the 378 class. *Sherman* requested, and was partially granted, an increase in manning for the deployment. The request was due to the increase in operational requirements for manning the combat information center, weapon postures, and for navigational requirements. So, the increased billets were in the quartermaster, radarman, and fire control technician ratings. We were also granted an additional health services technician (corpsman). So, our personnel allowance grew from 161 to 175 including an additional eight aviation personnel for the attached helicopter.

During the actual deployment we would end up losing twenty-five members. The losses were due to "A" school requirements, transfers, medical ailments, pregnancies, and discharges due to conduct. As non-rated members would leave to attend "A" school we would be sent replacements, and they arrived throughout the deployment. As we lost rated personnel they were not immediately replaced and we had to absorb their loss.

In preparation for deploying, in early April 2000 we departed for a two-month Alaskan fisheries patrol in the Bering Sea. On our return transit *Sherman* pulled into Seattle, Washington on the 14th of June to meet with the crew of the *USCGC Midgett (WHEC-726)*, the 378 that had just returned from an out of hemisphere deployment to the Persian Gulf. This was our opportunity to debrief the *Midgett* crew and get an idea of what we could expect. They provided us with many suggestions for making our deployment as smooth as possible. Surprisingly, their biggest complaint, and now our biggest concern, was the effects of the anthrax vaccination.

There was a major concern that Iraq would use chemical weapons such as anthrax against forces operating against them. Therefore, anyone who was getting ready to deploy to that part of the world was required to get the anthrax vaccination. The side effects had already been widely discussed in the other services, but hearing them first hand from these guys hit home. Most of their complaints were rashes and very sore limbs.

As luck would have it there was a shortage of the vaccine when we were preparing to deploy and we never received it.

In preparation for the deployment *Sherman* was scheduled to undergo a yard availability in Alameda, California and pre-deployment training in San Diego, California. On the 16th of June *Sherman* returned to Alameda and immediately commenced a dockside maintenance availability at the Coast Guard Island facility in Alameda. She then entered the Bay Ship & Yacht Shipyard in Alameda from 10 July to 1 September for major hull, structure, and machinery work.

***Sherman* at the shipyards in Alameda, California**

While at the shipyard the *Sherman's* Commanding Officer, Captain Dave Ryan, mustered the crew one day and asked us if we would be interested in circumnavigating the globe in conjunction with our out of hemisphere deployment. He asked for a sign of hands in favor. Since a majority of the crew was young, adventurous, and single, the vote was a majority "yes" for the circumnavigation.

The *Sherman* hummed with positive scuttlebutt as we commenced pre-deployment workups. There was some dissention and it seemed to be driven by a few of the crew's family members. They didn't want anything to do with being without their loved one for six months or more traipsing all over the world. In their minds the Coast Guard was to do as its name implied, guard the coast! They put up a good fight and even lobbied the Commandant of the Coast Guard in an effort to stop the deployment.

The crew's family members were very important to us. The captain had a liaison with the family members known as an ombudsman and she was Diane Ferguson, the wife of RDC Dennis Ferguson. And since this was such an important event, needing a lot more support and contact with family members, we had a co-ombudsman who was Renee Bowen, wife of EM3 Lyle Bowen. They would be an important part of our deployment, keeping information flowing back and forth from home to the cutter and producing a monthly newsletter. The ombudsman had a direct line to Captain Ryan, and he met regularly with Diane and Renee to discuss family related issues. This seemed to ease the stress the deployment was having on them.

On the 12th of October we commenced Phase I pre-deployment workups at the Navy base in San Diego, California which lasted a week. This consisted of classroom lectures for personnel manning the combat information center and was put together by Commander, Destroyer Squadron 23. The training provided a background on how current operations were being conducted in the Persian Gulf.

The 12th was also the day the *USS Cole (DDG-67)* was attacked in the Yemeni port of Aden killing 17 and wounding 39. Our crew watched the news coverage of the event unfold during class breaks and it brought a heightened sense of the dangers we were about to face.

On our way back to Alameda, the last week of October, we made a liberty port call at Catalina Island, California. This was an unusual port call for us because we had to anchor out in the harbor. Never the less the island was unique, a nice break, and well worth it.

Next, we stopped at the naval base in Port Hueneme, California where we spent a week conducting combat systems training and assessment. This was to ensure our fire control systems and weaponry were good to go. We also took advantage of the underway replenishment (UNREP) fuel simulator located there. This training would prove to be extremely important as we would UNREP ten times while deployed, taking on over 778,000 gallons of fuel.

Phase II workups were held between 30 November and 10 December back down in San Diego, CA. This consisted of detailed scenarios simulating routine Persian Gulf operations such as boardings, formation steaming, underway replenishment, gunnery exercises, and naval surface fire support. We conducted Phase II with the two U.S. Navy vessels that would deploy with us. They were the Spruance class destroyer *USS Paul F. Foster (DD-964),* which was around 529 feet long, and the Arleigh Burke class destroyer *USS Stethem (DDG-63),* which was about 500 feet long. This was not our first time meeting the *Stethem*. In early June 2000 she had attended the Rose Festival with us in Portland, Oregon.

USS Paul F. Foster **(DD-964) in the Arabian Sea**

The *USS Paul F. Foster (DD-964)* was commissioned on 21 February 1976 in honor of Vice Admiral Paul F. Foster, a Medal of Honor recipient for his actions at Vera Cruz, Mexico in 1914. *Foster* was no stranger to the Persian Gulf and had already deployed there six times, including participating in Operation Desert Shield and Desert Storm in 1990. This would be her 7th deployment to the Persian Gulf and her twelfth deployment overall.

USS *Stethem* (DDG-63) in the Arabian Sea

The *USS Stethem (DDG-63)* was commissioned on 21 October 1995 in honor of Steelworker Second Class (DV) Robert Stethem, a victim of terrorist atrocities during the high jacking of TWA 847 in 1985. This would be *Stethem's* third visit to the Persian Gulf and her third deployment overall. It is interesting to note that although she is a Navy vessel, she was awarded the Coast Guard Meritorious Unit Commendation Medal while on her first deployment for conducting search and rescue operations. The event occurred on the 23rd of November 1996 when she was diverted to help search for a downed U.S. Air Force C-130 aircraft, call sign "*King 56*" off the coast of northern California. Only one of the eleven crewmembers was found alive.

Both *Foster* and *Stethem* wanted to circumnavigate the globe with *Sherman,* but the request was denied due to a Navy requirement stipulating a certain port call quota. So they would have to return back the way they came. That meant *Sherman* would circumnavigate alone after departing the Persian Gulf.

We returned to Alameda in mid-December and focused on resting the crew. As the days grew closer for departure *Sherman* started her final preparations. On the 9th of January *Sherman* conducted a "fast cruise" in which all systems were checked for proper operation before getting underway. This consisted of starting and clutching in the main diesel engines, bow prop, setting the special sea detail, and operating radar. And on the 11th the aviation detachment (AVDET) from Air Station Kodiak embarked with their helicopter CG-6596.

We were ready to weigh anchor and set sail.

ENROUTE TO THE PERSIAN GULF

On the morning of the 13th of January 2001 the pier that *Sherman* was moored to at the Coast Guard base in Alameda, California came alive with family members gathering to wish their loved ones farewell. It was a crisp 53 degrees with a light breeze when our boss Coast Guard Pacific Area Commander, Vice Admiral Riutta, said a few words to the crew and family members. Afterward *Sherman* slipped her moorings at 0949 and made her typical transit out of the estuary and into San Francisco Bay, past Alcatraz Island, and under the Golden Gate Bridge. Only this time she was steering a course for the Persian Gulf, instead of her normal one to the Bearing Sea.

After departing Alameda *Sherman* plotted a course for Pearl Harbor, Hawaii at 15 knots and commenced Phase III workups. This consisted of working out connectivity issues on communication circuits with our Navy counterparts.

On the 17th of January we rendezvoused at sea with the *Stethem* and *Paul F. Foster*, the two much larger U.S. Navy ships that would make up our group for the transit to the Persian Gulf. *Sherman* would from this day forward always stick out as the odd white cutter amongst the fleet of grey Navy ships. Even though they were larger ships, we were commanded by a captain (O-6) and the Navy ships by commanders (O-5). This meant that *Sherman* was in charge of the three ship flotilla officially known as Surface Action Group (SAG) Pacific

Middle East Force (PACMEF 01-1). It also meant that we would lead the way to the gulf and at every port call have the choice berthing spot.

View from *Sherman's* deck leaving Coast Guard Base Alameda, California 13 January 2001

As the three of us headed across the Pacific Ocean the sea state was rough, with huge rolling swells, due to a big storm north of us. While transiting we spent our time standing watch and training in preparation for entering the Persian Gulf. There was always something to brush up on, even in rough weather. A majority of our training revolved around damage control drills.

Everyone on board is assigned a duty, or billet, for every type of damage control scenario. When the damage control alarm goes off it is known as going to "general quarters" and everyone is in a mad rush to get to their assigned billet. Cutters usually deploy to sea alone and have to be able to take care of themselves. So we train in everything from firefighting to a missile attack. Not only do we have to be proficient in taking care of ourselves, but we have to be able to provide assistance to other vessels in distress.

Before reaching Hawaii the sea state calmed down and on the 18th of January we were refueled at sea by the Navy oiler *USNS Yukon (T-AO-202)* and took on 35,614 gallons of fuel.

The next morning at 0817 on the 19th of January *Sherman* moored starboard side to pier B26 at Pearl Harbor, Hawaii. We moored at the Pearl Harbor Navy Base in-between the *USS Port Royal (CG-73)* and *USS Lake Erie (CG-70)*. Our moorings were just across the harbor from famous Ford Island where the *USS Arizona (BB-39)* rests. As with all naval vessels passing Ford Island and the *Arizona,* when we passed we manned the rails to honor all those who lost their lives on December 7th, 1941.

It was a great port call and most of us stocked up on personal supplies such as candy, Hawaiian coffee, and snacks. This would be the last major American facility we would be stopping at for almost 6 months. The two Navy ships loaded up with something a little more important, Tomahawk missiles.

On the 21st we left Pearl Harbor heading for the Coast Guard and Navy base at Guam. We steamed out of Pearl Harbor with the *USNS Yukon* and *SS Cape Girardo* along with the rest of our group. The *Yukon* and *Cape Girardo* sailed with us for about three days so we could conduct replenishing at sea and boarding team training with them. Once the training was completed on the 24th of January we refueled one more time with the *Yukon*, taking on 90,273 gallons of fuel, and they were released to head back to Pearl Harbor. As we headed towards Guam that day we crossed the 180th meridian on the 24th of January, which amazingly became the 25th of January!

There are several, centuries old, naval traditions that are observed to this day that revolve around sailing through certain areas of the seven seas. The open water is King Neptune's kingdom, and all sailors must pay respect to his highness. If not, you're sure to be sunk and join Davy Jones at the bottom of the sea! Crossing the 180th meridian, also known as the International Dateline, is one of those areas deemed sacred by King Neptune. When we crossed the 180th meridian on our Bering Sea patrols we always held a ceremony to appease King Neptune. It turned Polly Wogs, those who have never crossed the 180, into Golden Dragons, those who have. It does not matter how long you have been at sea or your rank; if you have never crossed you're initiated. So while steaming to Guam we initiated all Polly Wogs. Heck, the last thing we needed was to start this deployment out on the wrong side of King Neptune!

Meanwhile back in Alameda the ombudsman gathered family members at the Sizzler restaurant on the 27th to hold their first meet and greet since our departure.

While underway on the 28th we were able to watch Super Bowl XXXV thanks to satellite TV especially installed for this deployment. It may not seem like a big deal when you are used to having TV 24/7, but normally underway all we had were movies shown over the ship's entertainment system. So this was five star living! Well for the most part. We would soon find out that the reception had a lot to do with the weather and the course we were sailing, so maybe only 3 star living.

At 1444 on the 31st of January, we moored portside to wharf Victor at Apra Harbor, Guam alongside the *USCGC Galveston Island (WPB-1349)* who was homeported there.

The island of Guam is a tropical paradise. While walking to the beach I had the opportunity to tour some of the Japanese bunkers still intact from WWII. These bunkers had been part of the Japanese defense system and they dotted the islands coast. In fact almost 28 years after U.S. forces recaptured Guam in 1944 a Japanese Army sergeant was found still hiding in one!

We took advantage of the engines being secure by performing maintenance on them and had a bit of a personnel scare when an EM3 was shocked while working on a piece of electrical equipment. The shock was so severe that it caused his heart beat to get out of sync and he was taken to the local hospital for treatment. Thankfully he was returned later that day fit for duty.

After a brief stay in Guam we got back underway at 0833 on the 1st of February. Later that day at 1450 a seaman fell from a ladder in compartment 3-32-0-A, Bos'n Stores, and was unable to focus, think clearly, and complained of abdominal pain. At 1826 he was medevacked via CG-6596 to Anderson AFB and the naval hospital on Guam.

On the 3rd of February we crossed the Mariana Trench, also known as the Marianas Trench. The Mariana Trench is located in the western part of the Pacific Ocean and is the deepest water on earth, reaching a depth of 34,240 feet or 6.5 miles.

Of course we had to mark the occasion with a swim call held at roughly 11° 47.8N - 133° 58.2E.

Swim call at the Marianas Trench

As we transited through the Sulu Sea on the 6th of February we sailed through Philippine territorial waters making a course for Singapore. By the 7th we were in the South China Sea just north of Malaysia. We pulled into the military complex at Sembawang Wharves, Singapore and moored up starboard side to berth S-5 on the 9th of February. We had been underway 27 days now. The facility at Sembawang was home to the Command, Logistics Group, Western Pacific (COMLOG WESTPAC) and they supported the U.S. Navy's 7th fleet. It was a small facility and pier but big enough to squeeze us, the *Stethem*, and the *Foster* alongside in that order.

Stethem **being pushed into her berth behind** ***Sherman***

The complex also hosted units from the U.S. Air Force, Royal Australian Navy, Royal Navy, and New Zealand Defense Support Unit. There was also a nice club on base named the Terror Club. I could be wrong but I think that was an odd name for a club in that part of the world. The crew spent their time wisely, shopping and sightseeing. Access from the piers to downtown Singapore was easy. The locals relied on a very modern and impressive transit system known as Transit Link. It made getting around for us easy.

While downtown some of us visited The American Club which was exclusively for U.S. government personnel and their families. It was upscale and provided a safe environment to unwind.

Stethem **and** ***Paul F. Foster*** **moored at Sembawang Wharf**

This would also be our last pier-side port call before arriving in the gulf and our last opportunity to perform any maintenance. The facility at Sembawang had a contracting office to facilitate maintenance and repairs. The *Sherman* had a few minor items completed such as the captain's pantry refrigerator repaired and a chill water line installed. The *Stethem* and *Foster* had some bigger items completed such as resurfacing a deck.

At 0802 on the 15th of February we slipped our mooring and proceeded into the Johor Straight and it was on to Thailand. But first we needed to transit the Strait of Malacca. The transit can be tricky and dangerous. The strait is narrow, shallow, known for pirate activity, and can have limited visibility due to huge fires being burnt on the mainland. The fact that the later part of the strait was transited at night made for an even more challenging evolution. There were a large number of local fishing vessels operating in the strait with unconventional, limited, or no running lights! *Sherman* led the way steaming at 22 knots with *Foster* following behind at 3,000 yards and *Stethem* behind her at 3,000 yards. The transit was documented as being "fast but orderly."

In route to Thailand from Singapore, through the Strait of Malacca, with the *Paul F. Foster* and *Stethem* behind us. The big white domed piece of equipment is the 20mm CIWS.

We arrived in Phuket, Thailand at 1256 on the 16th of February and anchored out in Patong Bay along with our two Navy counterparts. The Navy really looked forward to their visits to Phuket. It was a regular stop for them on their way to and from the Persian Gulf. We had initially wanted to stop in Darwin,

Australia but the Navy didn't want anything to do with missing out on Thailand.

Some of the crewmembers took advantage of the once in a lifetime port call and had family members fly over to meet them. There was plenty to see and do in Phuket. From enjoying the world acclaimed beaches to riding elephants, it was a memorable port call! The family members that didn't make the trip to Thailand met at the Banner Club on Coast Guard Island back in Alameda for a pot luck dinner.

On the 22nd of February at 0947 after a long port call we weighed anchor. As we entered the Indian Ocean we set a course for the Persian Gulf at a speed of 15 knots. Time was not wasted during the transit. On the 23rd we conducted a .50 caliber gun exercise and helicopter operations. On the 24th we entered the Bay of Bengal and it was general quarters drills and more flight operations. It was also time to refuel and we took on 98,611 gallons of fuel from the *USNS Walter S Diehl (T-AO-193)*.

On the 25th we conducted more flight operations as we rounded India. As we neared Indian territorial waters we were hailed by the Indian patrol vessel *INS Sujata (P56)* and Indian aircraft. The 27th consisted of more flight operations to include landing the *Foster's* SH-60 and practice firing the 20mm CIWS. The CIWS was a very important weapon and our only protection from an incoming missile.

The *Sherman* and *Foster* each carried a helicopter on board. The *Sherman* had a HH-65A, call sign "Raven." She had newly fitted engines equipped with high performance combustion cans especially for this deployment. This type of helicopter is designed for search and rescue and did not have any type of weaponry or advanced night vision capabilities.

The *Foster* carried a SH-60B, call sign "Lonewolf." She was equipped with state of the art weaponry and surveillance equipment. This type of helicopter was designed for combat operations.

Owing to their significantly different helicopter designs the HH-65 was primarily used in day operations for boarding team security, passenger transfers, logistics, and support. And the SH-60 conducted almost all of the night time surveillance along with special warfare operations. Although limited in her special operations capabilities during Persian Gulf operations the HH-65 performed admirably and racked up over 143 flight hours and 79 sorties. Her performance during the entire deployment was over 227 flight hours and 131 sorties.

By the 28th we had entered the Arabian Sea. In preparation for entering the Persian Gulf we conducted more general quarters drills to include chemical, biological, and radiological (CBR).

USS Paul Foster, USCGC Sherman, and USS Stethem steaming in formation before entering the Persian Gulf

Before entering the Persian Gulf on the 2nd of March the Surface Action Group participated in anti-submarine operations in the Arabian Sea. The exercise scenario tasked a Los Angeles class submarine, the *USS Alexandria (SSN-757)*, with attempting to destroy *Sherman* while *Foster* and *Stethem* protected her. During the operation the *Foster's* helicopter was down due to a casualty, so *Sherman's* helicopter was used to deploy sonobuoys. This allowed *Foster* to detect the submarine before *Sherman* was destroyed.

It was now time to do what we had come here to do, board vessels. Being a boarding team member was a collateral duty. We didn't have personnel assigned to solely conduct boardings. Like most Coast Guard cutters our boarding teams were made up of the junior officers, cooks, quartermasters, radioman, and other ratings who volunteered to be boarding team members. And being on the boarding team did not relieve you of your primary job responsibilities. It was common for a team member to come back from a boarding and go straight to his or her primary job and finish out the work day or a watch.

Sherman had a pool of three boarding teams (black, blue, and gold) along with several backup boarding teams. Boardings were considered either compliant or non-compliant. The vessels we boarded were either happy to see us and compliant with our request to come aboard, or non-compliant and didn't want anything to do with us because they were up to something illegal.

Non-compliant boardings were unpredictable. Since the smugglers knew we were out there trying to stop them they welded their doors shut, welded bars over their port holes, and made any access into the inside of the vessel as difficult as possible. They also maneuvered their vessels erratically making the boarding process very dangerous. Their goal was to deter us long enough to get back into Iraqi or Iranian territorial waters where they were safe from us. Our goal was to get aboard, access the inside of the vessel, make our way to the pilot house, and take control of the vessel.

During our pre-deployment workups in preparation for non-compliant boardings *Sherman* sent fifteen members to be trained in tactics taught by U.S. Navy SEALs at their vessel boarding search and seizure (VBSS) facility in Hawaii. A majority of the training covered armed opposed boarding scenarios. This is the worst case scenario and trained our members how to deal with an armed threat. At VBSS this scenario was played out utilizing paintball guns to simulate real life gun fire.

Other aspects of training covered the inspection of cargo containers. Containers on commercial vessels are usually stacked several high and the boarding team member could find themselves needing to climb 30 to 40 feet to gain access. So, climbing techniques were taught on actual containers to give them realistic experience. The members also trained in the less dramatic aspects of a boarding such as documenting inspections, smuggler profiles, liquid cargo assessment, evidence collection, and file preparation.

The AVDET marking their helicopter, CG-6596, with our entry into the Persian Gulf

THE PERSIAN GULF

The Persian Gulf was a busy place and the *USS Harry S. Truman (CVN-75)*, an aircraft carrier, and her battle group were already operating there. *Truman's* battle group consisted of the *USS Stump DD-978, USS Deyo DD-989, USS Arleigh Burke DDG-51, USS Mitscher DDG-57, USS Porter DDG-78, USS San Jacinto CG-56, USS Carr FFG-52, USS Norfolk SSN-714,* and *USS Alexandria SSN-757*. Other U.S. and coalition ships operating in the gulf were the *USNS Pecos T-AO-197, USNS Kanawha T-AO-196, USS Ardent MCM-12, USS Cardinal MHC-60, USNS Catawba T-ATF-168, USNS Mount Baker (T-AE-34), RFA Orangeleaf A110,* and *HMCS Charlottetown FFH339*.

On the 16th of February in support of Operation Southern Watch *Truman* aircraft, along with other coalition aircraft, struck Iraqi air defense systems around Baghdad. This was in response to Iraq firing surface-to-air missiles (SAM) and anti-aircraft artillery (AAA) weapons on United Nation coalition aircraft. So, the atmosphere in the Persian Gulf region was very tense when we arrived.

On the 3rd of March *Sherman* manned her war time steaming configuration and set battle stations as we entered the notorious Strait of Hormuz for the eight-hour transit into the Persian Gulf. The *Stethem* was originally scheduled to lead as she was equipped with a state of the art Aegis weapon system.

But that system was not operational due to a casualty. So, *Sherman* once again led the way with the *Foster* and the *Stethem* following in that order. The Strait of Hormuz is the tiny entrance to the Persian Gulf connecting the Gulf of Oman with the Persian Gulf. The strait is lined by the country of Oman on one side and Iran on the other. To transit the strait, we had to enter waters claimed by each country.

Both the Iranian and Omani governments like to query U.S. Naval vessels and they hailed us on the radio as we steamed through with the *Foster* and *Stethem*. They want to know who is coming and going through the strait. The Commander of the 5th Fleet, who controls U.S. assets operating in the area, has "canned" responses to these queries. *Sherman* responded with the required response which was something like "This is United States warship 720 transiting." This threw the Iranian and Omani governments for a loop as they could not find a U.S. warship 720 in their catalogs! They continued to query us thinking we were an Ohio class submarine. When we finally identified ourselves as a U.S. Coast Guard Cutter that seemed to satisfy them, and we blew right through to the open waters of the Persian Gulf. After all the training hours spent in preparation to transit the Strait of Hormuz *Sherman* documented the evolution as "anti-climactic."

For the next two months *Sherman*, along with the *Foster* and *Stethem*, operated under the U.S. Navy's 5th Fleet. Our mission was to enforce the United Nation's sanctions against Iraq. The *Stethem* and *Foster* also had the added details of participating in Operation Southern Watch, Exercise Neon Falcon, Arabian Gauntlet, and carrier duties with the *Truman*.

The first thing a ship does when entering the Persian Gulf is refuel. On the 3rd of March *Sherman, Stethem*, and *Foster* lined up on the established fueling coordinates and took on fuel from the awaiting *USNS Pecos (T-AO-197)*. *Sherman* took on 117,155 gallons of fuel. We also took on 6 pallets of dry and frozen food stores via *Puma 202*, the *Pecos'* helicopter. This type of replenishment is referred to as Vertical Replenishment (VERTREP).

Getting ready to take on fuel from the *USNS Pecos* on her starboard side while *USS Paul F. Foster* refuels on her port side

Taking on supplies from *USNS Pecos*

During our two months in the Persian Gulf we would take on all of our fuel and stores while underway from *Pecos* and *Kanawha*. Taking on fuel and stores was an all hands evolution and we would do it about every 10 days. We refueled at sea three times on gas turbine propulsion and seven times on diesel engines. Both ways were effective. But being in charge of the engine room I thought the turbines sounded a lot cooler because when you went out on deck it sounded like a commercial jet taking off.

The *Sherman* pulling alongside of *Pecos* while *Stethem* approaches for refueling on the starboard side. Picture is taken from aboard *Foster* as she awaits her turn to fuel.

Pecos and *Kanawha* were busy and usually refueling ships on both sides, so we would have to get in line and wait our turn. When the ship ahead of us would break away we would steam up alongside and match the oiler's speed of 15 knots. Once that was accomplished we would pass over the required lines and take on fuel and stores. Our capacity was just over 200,000 gallons of fuel, so we would normally be taking on 50,000 to 90,000 gallons or so at a time. The evolution would last a few hours. Taking on fuel would only last about 15-20 minutes because the oiler sent it over at 800 gallons per minute! The majority of the evolution was taken up with getting into position and passing lines. When we finished we disconnected all lines and sped off.

All of our mail, fuel, food, and everything else we needed was either taken on underway from an oiler or delivered by the *Desert Duck 744* or *Mighty Duck 746* helicopters. These helicopters visited us every Tuesday and Saturday. They were older Navy HH-3 helicopters that looked like they had been worked to death over the years out here in the gulf. They would come zipping on up to us trailing a haze of black engine exhaust, hover over the flight deck, and drop off mail and small supplies. They would also take mail off and occasionally conduct passenger transfers. The mail the Ducks would bring out to us was extremely important to our morale.

The *"Mighty Duck"* helicopter dropping supplies off to us

The "*Mighty Duck*" making a drop

My roommate for most of the deployment was EMCS Joe Barthelemy. He ended up promoting to chief warrant officer while in the Persian Gulf and was flown off in the *Desert Duck 744* to his next assignment. He later told me that his ride on the Duck was one he would never forget. He said the door wouldn't close, fuel oil and hydraulic fluid were leaking all over the place, and he feared for his life!

After refueling at sea we headed to port for an operational in-brief from the Navy 5th Fleet. On the morning of the 4th of March we pulled into the Mina Sulman piers at Manama, Bahrain and moored up starboard side to berth 5. We would remain in port until the 7th of March. The Coast Guard had a port security unit (PSU) attached to the Navy 5th Fleet in

Manama, Bahrain. Among other duties they escorted vessels like us transiting the harbor.

Coast Guard Port Security Unit (PSU) escorting *Sherman* into Manama, Bahrain

The base at Manama, Bahrain was like an oasis. It had a little Navy Exchange store for essentials. There were also a few American fast food establishments, barber, education center, market place, and an open air bar which served alcohol, something not obtainable out on the economy. And we were allowed to go out off base onto the economy. Many of us took advantage of sightseeing and visiting the markets where gold jewelry was sold at a great price.

On the morning of the 7th of March we were underway and headed for the Northern Arabian Gulf ready to get to work.

Sherman's first assignment was to relieve the *USS Stump (DD-978)* that was providing watch over vessels being held at the Comiskey holding area, a designated area for vessels violating U.N. sanctions. Vessels were held there at sea until they were cleared or directed to a port for disposition. Our job was to make sure they stayed in the holding area and to physically board and check on them daily.

On the 8th of March after relieving the *Stump*, so she could head into Bahrain for a port call, we boarded our first vessels, the *M/V Asian Beauty* and the *M/V Amira,* sending over a boarding and the health and comfort team. I was the senior member of the health and comfort team and our job was to check for any health issues and to make sure the crew had enough food and basic essentials to last them for as long as we were holding them.

We were not at the Comiskey holding area for long. On the 9th of March the *Foster* relieved us and *Sherman* was tasked with working in the Northern Arabian Gulf to patrol for smugglers in an area known as Diamond Head. In route to that location at 0847 *Sherman* boarded the tanker *M/V Berge Ingerid*. She was found to be carrying legal cargo and released.

Our boarding process typically unfolded like this; a vessel was identified and intelligence gathered on it. Flight quarters was set and the helicopter launched for air support. The boarding team was notified and they would get themselves personally ready by donning coveralls, bullet proof vest, canteen, and personal effects such as toiletries. They would then check out their 9mm pistol along with ammunition and head to the flight deck, with the exception of the team lead who would report to the bridge for an intelligence brief.

As the boarding team assembled on the flight deck they would man the rail. Pointing their weapon out to sea they would insert a loaded magazine and chamber a round. From there they would enter the small boat which was situated on the side of the flight deck and be lowered to the water. Once in the water the team was whisked over to the suspect vessel.

Boarding team climbing aboard a suspect vessel

Once in the Northern Arabian Gulf numerous vessels were monitored traveling down the Khwar Abd Allah River, into Shatt al-Arab, Iraq and transiting along Iraqi and Iranian territorial waters, obviously smuggling oil. *Sherman* tracked a dozen non-compliant vessels as they snuck through the straight and into open water. But each time we prepared to board them we were ordered to break off contact.

Boarding team departing a suspect vessel

This was documented by *Sherman* as being very frustrating. Non-compliant boardings were dangerous and the operational commander, *Sherman's* boss, had withdrawn our authority to conduct them. The only unit allowed to conduct them were the Navy SEALs.

Captain Ryan was determined to get into the action. He lobbied the operational commander and negotiated our authority to conduct non-compliant boardings. The smugglers seemed to have caught wind of *Sherman* getting into the game and slowed their push out of Shatt al-Arab.

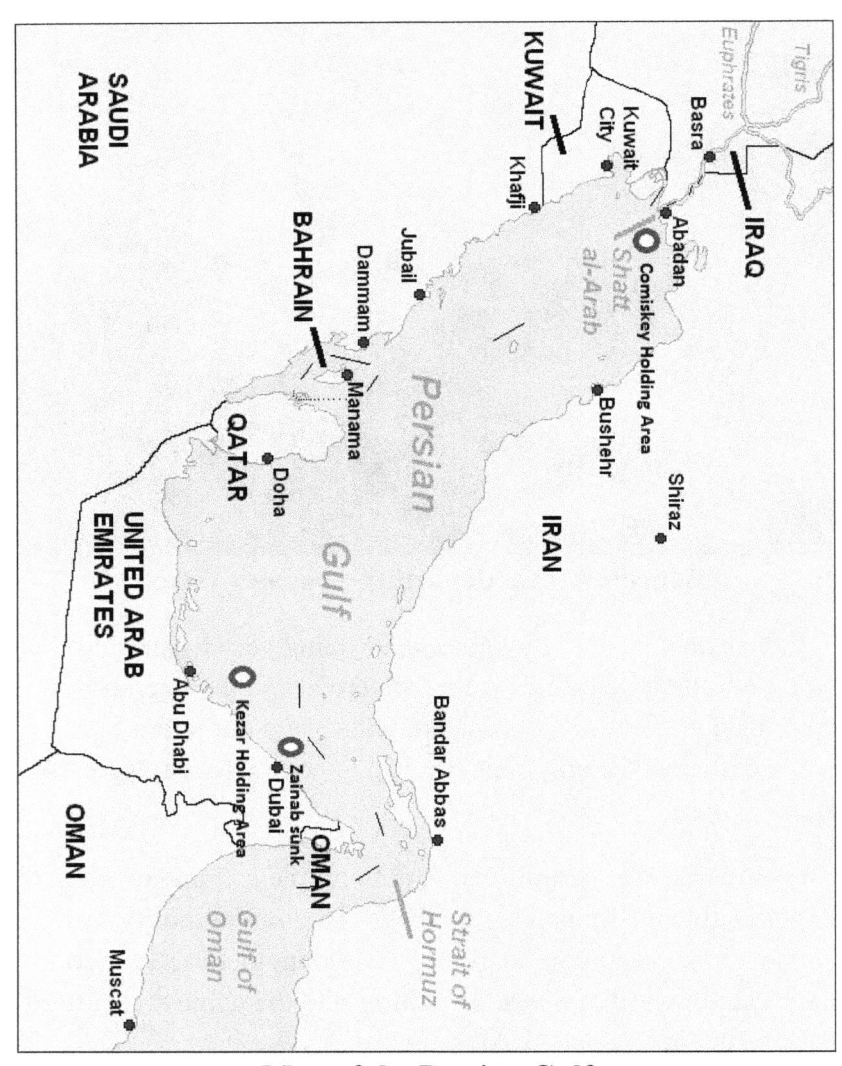

Map of the Persian Gulf

On the 10th we had our first visit from the *Desert Duck 744*.

By the 12th we needed more fuel. Between gas turbine operations and flying the helicopter several times a day we were going through it. So at 1504 we lined up on the *Kanawha* and took on 41,988 gallons of fuel.

Sherman headed for a port call on the 13th of March and moored starboard side to berth S-5 in Bahrain once again. Across the pier from us was the Bahrainian Navy frigate *RBNS Sabha (FFG-90),* which was given to them by the United States in 1996. When a U.S. asset she was the *USS Jack Williams (FFG-24).* If there was ever any doubt as to where we were, it was clearly evident when the *Sabha* piped "call to prayer" several times a day and the crew would stop what they were doing and prayed.

While in port Manama, Bahrain I was visited by the Navy's 5th Fleet Command Master Chief, MCPO Terry Scott. Each time we pulled in he would be waiting on the pier to come aboard and see me to discuss any enlisted issues we had. He was always concerned about how we were getting along and if there was anything the 5th Fleet could do for us. He was very professional and I always felt he was truly concerned about our needs. I was not surprised when he later became the 10th Master Chief Petty Officer of the Navy (MCPON).

During this port call I also had the added pleasure of having the Reserve Force Master Chief of the Coast Guard, MCPO George Ingraham, pay a visit. He was stationed at Coast Guard Headquarters in Washington, DC. and visiting the reserve forces deployed in Bahrain.

In *Sherman's* Chiefs Mess (left to right) Master Chief George Ingraham, Senior Chief Meyers, Senior Chief Widman, Master Chief Terry Scott, Master Chief Semler.

I also had the privilege to tour a U.S. Navy Los Angeles class nuclear submarine, the *USS Alexandria (SSN-757)*. She was the submarine we conducted anti-submarine operations with prior to entering the Persian Gulf. She was moored across the pier from us and I walked over to invite their chief's mess over to the *Sherman* for introductions. While there, the *Alexandria's* command master chief took me on a tour. The entrance to the sub was a small vertical shaft that was a bit tricky to get down. From there all I can say was that it was unbelievably cramped and not for the claustrophobic! After the tour I had a much better appreciation for my accommodations on *Sherman*.

For most of March while *Sherman* conducted boardings the *Stethem* and *Foster* along with the *Truman* Battle Group participated in Arabian Gauntlet. This was an eleven nation exercise including more than twenty ships and lasting twenty-one days until the 1st of April. During this time *Sherman* was given the privilege of assuming the duties as Commander of Maritime Interdiction Operations.

It was under way again on the 16th and we headed into the Northern Arabian Gulf. We were once again patrolling around the Comiskey holding area just off the coast and slipping in and out of Iranian territorial water. We were very busy! On the 17th at 0440 we started to track the Syrian flagged *M/V Ampoz* which we escorted to an anchorage point. At 1018 we were tracking the *M/V Al Juzuf* and at 1050 the Iranian flagged *M/V Senobar*. At 1843 we made contact with the Chinese flagged *M/V Luban* and at 2340 we identified the Sri Lankan *M/V Sigiri*. These vessels were all tracked but not boarded.

18 March at 0603 we entered Iranian claimed territorial waters trying to intercept the 160-foot *M/V Al Maha*. Captain Ryan ordered our watch officer to direct the *Al Maha* to turn around. Meanwhile to prove we meant business the 76mm and 25mm guns were turned and pointed at the *Al Maha*. Once again Captain Ryan ordered the *Al Maha* to head south out of Iranian claimed waters. The *Al Maha* replied "general cargo, general cargo." Captain Ryan once again instructed our watch officer to inform the *Al Maha* that we will notify Iranian officials and the nearest Iranian port of their presence. The *Al Maha* responded by making a sharp turn towards *Sherman* causing us to divert out of her path. The *Al Maha* then turned and slipped deeper into Iranian claimed waters and out of our reach. We eventually

broke off pursuit at 0649. Slipping in and out of Iranian claimed waters wasn't unusual for us as we pursued suspect vessels. Just like pointing our weapons at her, it was all part of the game of cat and mouse. In this case the *Al Maha* was too far into Iranian waters for us to board. There was some good news though, the *Desert Duck 744* made a call and delivered 27 pounds of mail!

Sometimes we were literally juggling contacts. On the 19th of March at 1927 we steamed close to the *M/V Diamond* which was under Honduran flag. It refused to respond to radio transmission and we were directed to break off contact at 1941. That was fine because we immediately came upon the *M/V Agon* at 1957. After establishing contact with her she was directed to an anchorage point. This was a busy area and at 2236 we made contact with the *M/V Hanen*. After establishing contact she was directed to the same anchorage as the *Agon*.

Things were broken up a bit on the 20th. We had been trailing an unidentified vessel but broke off contact due to a medical emergency with an RD3. It was determined that the member would be flown to the *Truman* for treatment so we set a course for her. When we were within helicopter range we set flight quarters and the RD3 was transferred to the *Truman*. His condition was so severe he was sent back to Alameda, California for further treatment.

This medical issue is probably the strangest one you will hear about happening aboard a ship. The RD3 must have unknowingly picked up a poisonous spider before we entered the Persian Gulf, probably in Guam, Thailand, or Singapore. He reported what appeared to be a classic spider bite in Singapore and received surgical treatment there. Shortly before reaching

Bahrain he received another bite. He was taken to the hospital in Bahrain and upon evaluation found to have multiple bites. And now he had more bites requiring him to be medevacked to *Truman*.

Well obviously if he was bitten and sick while underway we had a poisonous spider aboard! The members berthing area was searched, the spider was found in the members rack, and then killed.

On the 21st it was back to checking vessels and we queried the *M/V Karem* as she steamed along at 9 knots. Our normal operating mode for querying vessels was to idle on station or slow patrol on one diesel engine to conserve fuel. When a suspect vessel was identified we would come up on the second diesel engine or turbines depending on how fast we wanted to close the gap.

On the 22nd at 1115 we maneuvered to intercept an unknown vessel travelling at 5 knots. At 1150 we made contact with the Sao Tome flagged *M/V Hassan 1*. The vessel was released. At 1433 we met up with the *Foster* and Captain Ryan along with Commander Klipp transited via small boat to her for an intelligence brief lasting about two hours.

The morning of the 23rd at 0232 we had two Navy Seal small boats tie up alongside of us for a few hours and then depart on a special ops mission. A few hours later at 0443 we identified the Belize flagged *M/V Yemaya*. The helicopter was launched for security and surveillance and eight members of boarding team blue were lowered into the small boat. At 0700 boarding team blue was aboard the *Yemaya* with CG-6596 providing air support. The boarding was completed at 1159 and the *Yemaya* released. Later that afternoon we made contact with the *M/V Gem* traveling at 5 knots and the *M/V Al Debran* who was speeding along at 9.5 knots. We shadowed them until they both crossed into Iranian waters and we lost radar contact.

We had burned a lot of fuel over the past twelve days and it was time to replenish. On the 24th we lined up on the *Pecos* and took on 53,993 gallons of fuel and sixteen pallets of stores.

By the 25th *Sherman* had drifted over to an area around the island of Khark, which is just north of the city of Bushehr, Iran and made contact with the *M/V Ala Bas*. By the time boarding team black got aboard her she was in Iranian recognized territorial waters. Never the less the boarding was completed and the vessel released.

26 March yielded a double boarding. At 0325 *Sherman* commenced tracking *M/V Swissco* in Iranian claimed waters and placed boarding team gold on her at 0851. At the same time the *M/V Yick Lee* was also boarded by boarding team blue. No violations were found on either vessel and they were released. While the boardings were taken place the Commanding Officer of the *Foster* came over to *Sherman* via small boat for an intelligence and operations brief for an hour and a half.

27 March was yet again another busy boarding day. We were operating with the *Foster* and at 0725 boarding team gold was placed aboard the *M/V Tamatiki*. This would be a long boarding lasting until 1803. As this was a Tuesday, while conducting the boarding, the *Desert Duck 774* payed us a visit.

At 0707 the following day, the 28th, the gold boarding team was once again sent over to the *Tamatiki*. She wouldn't be released until 1749 that day.

On the 29th the gold team was up again at 1211 and they boarded the *M/V Jordan II*. At 1302 the *Foster's* small boat came alongside *Sherman* and picked up the blue boarding team and transited them to the *M/V Al Debran*. We had broken off contact with the *Al Debran* back on the 23rd after she crossed into Iranian waters but got another crack at her today. At 1540 the gold team was back on board *Sherman* and at 1722 the blue team arrived back aboard.

Sherman had maneuvered close to Shatt al-Arab, Iraq on the 30th and was moving to intercept the *M/V Bink Star* at 0420. In the meantime we ran into the *M/V Petronia* and placed the busy gold team on her at 0844. It was another long day and the gold team was back on board at 1847.

The 31st had *Sherman* maneuvering to intercept a vessel at 0331 that was putting along at 4 knots but lost contact. Later we made contact with a vessel at anchor. It was the nine-member blue boarding teams turn and they made their way aboard the *M/V Ibn Hazm* at 0845. This was a Saturday and while the boarding team was over on the *Ibn Hazm* the *Desert Duck 744* and *Sherman* transferred mail. At 1338 Captain Ryan departed for the *Ibn Hazm* to have a look at her and returned to *Sherman* at 1427. At 1737 the boarding team was back on board *Sherman* and the *Ibn Hazm* released.

Things quieted down on the 1st of April as we patrolled along with *Foster* in the Comiskey holding area.

On the 2nd we were back at it and had stopped the *M/V Caria* and placed our climbing team aboard at 1513. They were specially trained to climb shipping containers looking for contraband. At 1923 they were back aboard and the *Caria* was released.

The 3rd was Tuesday and arrived with *Sherman* sitting dead in the water looking for radar contacts in the middle of the gulf, off of Kuwait City. Besides launching the helicopter to conduct search grids the only notable event was the *Mighty Duck 746* coming out for a mail transfer as scheduled.

As we prepared for an upcoming port call on the 4th we took the time to hold a gunnery exercise with the CIWS and 25mm guns. We also lined up on the *Kanawha* once again and took on 54,195 gallons of fuel.

On the 5th of April we made our way back into Bahrain and moored starboard side to berth 10. We would stand down from the 5th to the 10th of April.

It was underway again on the 10th of April and we had the opportunity to host a visit by our Coast Guard boss, Vice Admiral Riutta. We then coordinated a visit for him with our Navy boss aboard *Truman* and a visit with the captain and crew of the *Pecos*.

Things had cooled off in the Northern Arabian Gulf and *Sherman* was re-assigned to the Southern Arabian Gulf (SAG) and the Gulf of Oman. While most of the U.S. and allied warships were participating in another joint exercise, Neon Falcon, *Sherman* was on the prowl for smugglers.

This is when things heated up!

Stethem needed to depart for carrier duties with the *Truman* in support of Neon Flacon and handed *Sherman* custody of the *M/V Diamond* which was anchored in the Kezar holding area. This holding area was just off the coast of the United Arab Emirates (U.A.E.)

While maintaining custody of the *Diamond* I boarded her several times along with one of our corpsman, HSC Liz Beck, conducting health and comfort visits. *Sherman* also maintained a six member custody crew aboard *Diamond* which was rotated every 12 hours. Eventually *Sherman* was tasked with escorting *Diamond* to Abu Dhabi, U.A.E. for disposition.

On the 11th of April while in transit with *Diamond, Sherman* was redirected to rendezvous with *Foster* in the Northern Arabian Gulf to take over custody of the Georgian flagged *M/V Zainab* enabling *Foster* to proceed to Neon Falcon. *Sherman* left a six man custody crew with *Diamond* and steamed to meet *Foster*. After taking custody of *Zainab, Sherman* placed a custody crew aboard her. The *USNS Catawba (T-ATF-168)* aka *"the Desert Cat"* rendezvoused with *Sherman* and continued to escort *Zainab* along with *Sherman* back to secure *Diamond*.

While in transit on the 12th the *Mighty Duck 746* delivered four Navy EOD divers and four loads of cargo and mail. *Sherman* was experiencing severe vibration with her propulsion shafting and the divers were flown out to inspect them. It was an interesting evolution with the *Mighty Duck* lowering the divers and their gear down on to our flight deck. We anchored in 208 feet of water so the divers could clear fish netting from our propeller shafting. When they finished the *Mighty Duck* came back out and hoisted them and their gear back up and off they went. With the fish netting cleared *Sherman* and the flotilla proceeded on.

By the 13th *Sherman* had four gold boarding team members on *Catawba*, seven gold members on *Zainab*, and eight blue team members on *Diamond*. When the *Zainab* was caught she was loaded down with over 3,800 tons of illegal Iraqi oil and was headed out of the Persian Gulf to Pakistan. The Navy boarding team that sized her had been told by *Zainab's* crew that her only two diesel fuel storage tanks were contaminated with seawater. Therefore the Navy crew was having to transfer diesel fuel via a small boat in 5 gallon cans to *Zainab* to keep the engines running, a very labor intensive process. When *Sherman* took custody our Engineering officer, LT Dave Socci, went aboard to investigate and within hours found a third diesel fuel storage tank full and uncontaminated. The crew had been lying about not having a good fuel source.

Zainab and her crew were bad characters.

In September 2000 she was intercepted by *HMS Marlborough* but released as she was unseaworthy and sabotaged. In December 2000 she was intercepted by the U.A.E. In January 2001 she was intercepted by *USS Hewitt* and released after being cleared of an engine casualty. And finally in February she was intercepted by *USS Arleigh Burke,* but released once again for being unseaworthy.

On the 14th of April at 0702, while transiting back to *Diamond,* along with *Zainab* and *Catawb*a, *Sherman* conducted a morning boarding on *M/V Qamar*. *Sherman* had no problem multi-tasking! The vessel was cleared as having legitimate cargo.

Zainab and *Catawba* were slowly trailing behind *Sherman*. They had dropped behind as they had to transit around several Iranian claimed islands and the sea state had worsened. Unknown to the custody crew, the captain of the *Zainab* had been pumping sea water into the ship thinking we would release his vessel if we thought it was sinking and unseaworthy as had happened in the past.

The *M/V Zainab* taking on water

At 0815 *Sherman* received a report from *Catawba* that *Zainab* was taking on water. By 0850 *Sherman* had the rescue and assistance team aboard *Zainab* conducting damage control. MKC John Young was aboard Zainab and told me the crew acted indifferent about the condition of their vessel, almost as if they had been on a sinking vessel before.

As I watched from the *Sherman* you could see she was very low in the water and was producing a large sheen of oil. As *Zainab* steamed closer she was taking on water over her port side and this forced her in a circular motion to port as if her rudder was hard over. The rescue and assistance team upon investigating below decks determined the vessel was sabotaged and flooding. They quickly set up dewatering measures to slow the flooding. By this time she had drifted into U.A.E. waters and they had sent a small boat out, *U.A.E. 655*, to investigate. At 1247 the eleven member crew of *Zainab* were transferred to the *U.A.E. 655*. At 1259 six of the boarding team members were removed and transferred back to *Sherman*. As the small boats circled *Zainab* and transported personnel to and from her their hulls became coated with floating oil. From the deck of *Sherman* the smell of crude oil permeated the air.

M/V *Zainab* taking on water

The rescue and assistance and boarding team kept trying to dewater the vessel but the flooding was too bad and it finally took the vessel over. At 1322 when all hope of salvage was lost Captain Ryan gave the order to pull the remaining team members from the vessel. Minutes later at 1340 we watched the *Zainab* roll to her port side and her bow start to go under.

Evacuating the boarding and rescue and assistance teams from *M/V Zainab*

It didn't take long for the *Zainab* to sink like a rock. After she went down she left behind a pretty good sized debris field of anything that would float. To limit the navigational hazard this floating junk may cause, *Sherman* opened up on the junk with machine gun fire in an effort to sink the bigger items. After sinking a majority of the floating hazards *Sherman* secured machine gun fire due to vessel traffic in the area.

I will have to say that over my eleven years at sea I have responded to countless search and rescue cases involving vessels sinking, but I have never witnessed a vessel of this size going under. Watching the video I took of the event years later, it's still amazing!

M/V Zainab **sinking**

The *Zainab* is now a permanent fixture at the bottom of the Persian Gulf under about 85 feet of water, 20 miles off the coast of Jebel Ali, U.A.E. at 25° 14.9'N 054° 51.5'E.

We lost the *Zainab* but we still had the *Diamond* in custody.

On the 15th I boarded *Diamond* and had an interesting encounter with their engineering officer who spoke broken English. I was conducting the boarding to find out about liquid load information which he provided me. We then sat down to eat the military's prepackaged meals ready to eat (MRE) together. As a good will gesture while conducting health and comfort boardings we always tried to get to know the crew and would break bread with them. He and his crew seemed to really enjoy the small bottles of hot sauce in the MREs and he asked if I had more. I really didn't like hot sauce and had a bunch of left over bottles in my back pack and gave them to him.

He was an interesting Iraqi man who was obsessed with the cost of goods in America compared to Iraq. In our conversation he was determined to convince me that Iraqi goods were cheaper than goods in the United States. He seemed to have pleased himself when he asked me how much a chicken costs in the United States, and I gave him my answer of about $3.00. He said in a proud voice that it was much cheaper in Iraq! With all the sanctions and turmoil going on in his country I wondered if you could even find a chicken in Iraq.

I would spend the night on *Diamond* with the rest of the boarding team as we spent the 16th transiting. On the 17th we had eight boarding team members on *Catawba* and seven on *Diamond*. We were using the *Catawba* as a platform to stage boarding teams to rotate out on *Diamond* so we could still conduct flight operations and vessel surveillance. The three of us were slowly making our way to Abu Dhabi, U.A.E. for disposition.

By this time the captain of the *Diamond* had lost his patience with us and submitted a letter demanding we release his vessel. Captain Ryan replied that his letter would be forwarded to the proper authority, but as long as his vessel was in Coast Guard custody the *Diamond's* captain was responsible for the safety of his vessel and crew.

The 18th found us still making our way to Abu Dhabi. The *Diamond* was experiencing an oil leak on one of its engines. I went back over along with several other machinery technicians and we stabilized the oil leak. After getting back on board *Sherman* we left the *Diamond,* with twelve members of security team "Q" aboard, and *Catawba* to continue on to Abu Dhabi and we steamed off and lined up on *Pecos* to take on 90,521 gallons of fuel and 26 pallets of cargo via their helicopter.

Refueled and resupplied, Captain Ryan requested clearance to conduct intercept operations in the Southern Arabian Gulf and Gulf of Oman with hopes of locating the high value smuggler *M/V Fal II*. This was granted and on the 19th in the Gulf of Oman we started to query the *M/V Al Asshaar*. On the 20th we started tracking the Honduran flagged *M/V Hurmoz* in relatively the same location as *Asshaar*.

The 22nd had family members gathering back in Alameda for an ombudsman sponsored picnic. In the gulf we continued patrolling and had moved into the Strait of Hormuz. At 0923 on the 22nd in location 26° 24.1N 56° 43.6E we made contact with the *M/V Georgios* and maneuvered to intercept her.

When we neared the strait an Iranian Hendijan class gunboat come out to make sure we stayed out of their territorial waters as *Georgios* steamed just inside of them. The gunboat came right up to us in a threatening manner and we set general quarters. As they shadowed us we saw that they had their guns covered, so they posed no immediate threat. Never the less we stayed prepared and they shadowed us until we reached the strait, at which point they broke off. This class of vessel is known to carry Noor missiles. But after reviewing my video of the event I could not see them mounted on the vessel that came out to inspect us.

Iranian Hendijan class gunboat shadowing us

When we cleared the Strait of Hormuz there was a window of opportunity when *Georgios* was in open water and fair game. That's when we had boarding team black ready and waiting to board her. As the boarding team approached the captain of the *Georgios* threatened to blow his vessel up if we attempted to board her!

As the boarding team stood off of *Georgios* they noticed that although the captain was calling his vessel *Georgios,* the name painted on the stern was *Gillian Everard, London.* The *Georgios* was still acting erratically and threated to ram the small boat that the boarding team was in. At this point our window of opportunity to seize *Georgios* was closing and Captain Ryan decided to break off contact with her and she once again entered Iranian waters.

We continued to track *Georgios* and on the 24th at 1707 we intercepted her in the middle of the Gulf of Oman north of Muscat, U.A.E. She was just over 260 feet long and was designed to carry cargo, not liquid loads. Her pilot house sat to the rear of the vessel and she had two cargo masts, one amid ship and one toward the bow.

Boarding team black was readied and sent to board her with *Sherman* standing off for protection. Team black made a quick board at 1745 and the *Georgios* captain was reported as being compliant, although he claimed that the vessels steering was inoperative. I was sent over at 1812 to assist the boarding team and investigate the steering casualty. At 1845 boarding team golf with CWO Sepp in charge was onboard *Georgios*. *Georgios* was found to be carrying illegal cargo. At 1946 the black team was relieved and sent back to *Sherman*.

CWO Paul Sepp smiling on the bridge wing of the *M/V Georgios*

During this event we also detected a second suspect vessel, the *M/V Kade Jah* transiting along the edge of Iranian territorial water. On the 25th we intercepted her and boarding team black was placed on her at 0504. We thought she looked familiar. It turns out she was recently renamed and was previously the *M/V Al Maha*. That's the name of the vessel we tracked and had to let go back on the 18th of March after she slipped deep into Iranian waters. Needless to say this time she was found to be carrying illegal cargo.

We still had control of *Georgios* and at 1726 we exchanged boarding teams so the gold team and I were back on *Sherman*.

With the *Georgios* and *Kade Jah* in custody we headed back into the Persian Gulf. On the 26th security team "Q" caught back up to us! MK1 Andy Vandewarker, who was part of the team, said they had turned over custody of the *Diamond* to the U.A.E. Coast Guard and transited back to Bahrain with the *Catawba*. They stayed in Bahrain a few days and then were flown out on the *Desert Duck 744* to the *USNS Mount Baker (T-AE-34)* and from there transferred to *Sherman* via small boat. The *Mount Baker's* helicopter, 4824, also transferred 3 loads of mail to us.

At 0822 on the 28th we handed them over to the *USS Mitscher DDG-57* at the Kezar anchorage.

Not long after handing the *Georgios* over to the *Mitscher* she was once again causing problems. The *Mitscher* was acting as the guard ship over her on the 30th of April when two of her crewmembers dove into the snake infested waters of the Persian Gulf. The *Mitscher* lowered her RHI small boat to retrieve the two crewmen. Initially they put up a fight but *Mitscher* noted "after they realized their peril" gave up and were eventually hauled out of the water and returned to the *Georgios*.

A few days later *Mitscher* handed off responsibility of the *Georgios* to *HMCS Charlottetown*. But this would not be the last time the *Georgios* would be caught smuggling oil. Four months later, in August, she was boarded by a U.S. Navy boarding team in the Northern Arabian Gulf. After boarding the *Georgios* the boarding team noticed an oil sheen coming from her and she was determined to be sinking, more than likely sabotaged. The twelve member Iraqi crew was evacuated, two to the *USS Enterprise (CVN-65)* due to pre-existing medical conditions and the rest to the *USS Stout (DDG-55)*. The *Georgios* subsequently sank about 60 miles off the coast of Kuwait with an estimated 2,898 tons of illegal oil.

In the midst of everything going on with the *M/V Zainab, M/V Georgios* and *M/V Kade Jah* coalition aircraft, including those aboard the *USS Truman,* had targeted an Iraqi anti-aircraft site on the 12th of April, a mobile early warning radar site on the 19th of April, and a radar & artillery site in Najaf providence, about 100 miles south of Baghdad on the 28th of April. The sites were targeted to protect coalition aircraft enforcing United Nations mandates.

Later in the day on the 28th after being relieved of the *Georgios* and *Kade Jah* we lined up on the *Pecos* one last time and took on 92,600 gallons of fuel.

On the 29th we pulled into Manama, Bahrain for six hours to handle logistics issues and debrief 5th Fleet. At 1810 we slipped our berth and started to slowly head for the Strait of Hormuz and our exit from Persian Gulf operations.

Sherman ended up conducting 219 queries, 115 boardings, and 5 diverts of vessels. The *Stethem* recorded the third largest arrest of a vessel violating oil sanctions when she seized the *M/V Diamond* carrying 7,462 tons of illegal oil. The seizure also included documents that provided intelligence on illegal vessel trafficking routes. *Stethem* in their command history report stated "The seizure of this oil prevented Saddam Hussein from making two million dollars."

Once again on the 30th of April we transited the Strait of Hormuz and started the two month trip back home. The *Stethem* and *Paul F. Foster* escorted the *USS Harry S. Truman,* who was on her first deployment, through the Strait of Hormuz earlier on the 27th of April. Then they headed back home. Of course, instead of going back the way we came, we continued on our journey around the world, and plotted a course for the African continent.

As we entered the Gulf of Oman late on the 1st of May we arranged to pick up malaria medication from the *USS Constellation (CV-64)*. She was making her 20th deployment and just entering the Persian Gulf with her battle group, relieving the *Truman*. As preparations were made to fly our helo over to *Constellation* one of our crewmembers mentioned that his brother was stationed on the *Constellation*. It was arranged so that our crewmember, SA David McCalister, could fly over with the helo and catch up with his older brother AE3 Joseph McCalister.

AE3 Joseph McCalister with his brother SA David McCalister onboard *USS Constellation (CV-64)*

HEADING HOME TO ALAMEDA, CALIFORNIA

The day after steaming out of the straits you could feel a weight lifted off of the crew. Two months of war time steaming and boardings had taken its toll on us. We had felt like we were constantly at work. About the only break was hitting the rack for a few hours of sleep when we could. Now we were getting back to the normalcy we knew while being underway.

On the 5th of May *Sherman* crossed the equator heading south at 56° 15.0E. In recognition of the equatorial crossing King Neptune visited us once again as all "Polly Wogs" became "Shellbacks." Like all line crossings *Sherman* conducted a ceremony and a swim call was held.

During the transit south we also conducted helicopter operations and gunnery exercises to keep us sharp. Now we were off for some much needed port calls and liberty. We were planning a stop in Tanzania but that was scrubbed from the itinerary due to civil unrest in the country. Fine by us, we had enough drama the past two months!

From the 6th to the 10th of May we made our first port call in Victoria, Seychelles mooring starboard side to Mahe Quay. Seychelles is made up of tiny islands located off the east coast of Africa and has a population of just over 90,000 people. There wasn't really much going on when we visited and the town seemed to be shut down or on some sort of holiday. It was a nice relaxing way to break ourselves back into going on liberty! I spent most of my time walking around the island sightseeing. The downtown area seemed to be centered on a huge cricket pitch and I assumed that was the local favorite past time.

As with all port calls there are always crewmembers on duty 24 hours a day standing watch, taking care of refueling, bringing on stores, and performing maintenance. The duty section on one particular day had local divers come out to inspect the underside of the cutter. Like in the gulf they came up with huge amounts of fish netting that they said were wrapped around the starboard propeller shaft.

The AVDET guys were also kept busy in port and conducted a helicopter rescue swimmer and basket hoist demonstration with one of the Seychelles Coast Guard vessels. This vessel turned out to be a 44-foot MLB recently given to them by the U.S. Coast Guard!

The crew also participated in a community relations endeavor to clean up Cousin Island. It was located about an hour and a half boat ride away from the main island of Mahe. The work crew consisted of nineteen crewmembers.

***Sherman* moored in Victoria, Seychelles**

We left the Seychelles and after only two days of transit pulled into our next port of call, Port Louis, Mauritius on the 13th of May. This island country was much livelier then Seychelles and bustling with activity. Tours were set up for the crew and the pace of liberty was stepped up! I personally visited an inactive volcano, numerous breathtaking water falls including the famous Chamarel Falls, and the Seven Coloured Earth, also at Chamarel.

As in several of our port visits we hosted a formal meet and greet party on the cutter for local dignitaries and embassy personnel. These were held on the flight deck and enabled our guests to enjoy an evening of socializing on the cutter.

Social event on the flight deck in Mauritius

On the evening of the 16th the officers and chiefs aboard *Sherman* were invited to the house of Charlie and Lizzie Slater who worked for the U.S. Embassy. The Slaters had a long history of service in Africa, which almost came to an end in August of 1998 when the embassy in Dar es Salaam, Tanzania and Nairobi, Kenya were bombed by terrorist on the same day killing over 240 people and injuring over 4000. Elizabeth Mary "Lizzie" Slater was only several days into a new position at the embassy in Dar es Salaam, Tanzania when she was severely injured. Charlie Slater worked several hours away at the embassy in Nairobi, Kenya was uninjured. They both stayed on at their prospective embassy positions until they transferred to Mauritius in 2000.

Invitation to the Slater's

They were great hosts. It was a memorable evening punctuated with the chef preparing this huge dish of paella. He was cooking it over an open wood fire in a huge 3 or 4 foot round pan. I had never had it before and it was delicious!

The crew also had the opportunity to give back to the people of Mauritius for their hospitality and spent time painting a local boys' school and delivering books and medical supplies to local schools.

On the 17th we weighed anchor from Mauritius and headed to Madagascar arriving on the 19th of May in the northern port town of Antsiranana, Madagascar (known as Diego Suarez prior to 1975) and moored port side to the city pier. The people of Antsiranana had not welcomed a United States ship since 1984 when a Navy vessel deployed there to provide aid after a cyclone.

Madagascar was the 4th poorest country in the world at that time and it looked it! The people were fabulous but the infrastructure was a total mess. The people were so nice but had nothing. They obviously didn't even have basic garbage service because everyone just dumped it in any vacant space they could find. Goats and chickens roamed around the streets.

Part of our visit was to provide humanitarian supplies, mostly medical, to the local community. We didn't have much but every little bit helped. The crew carried several hundred pounds of rice, books, medicine, and toys to the local hospital, children's daycare, orphanage, hospice, and to a women's organization.

Probably the biggest event for the crew was a soccer game between our crew and a local team. It was a big event for the community and covered by the media. In attendance were several dignitaries including the president of the province, the minister of defense, the general in charge of the regions military, and the Madagascar Naval Commandant. For just a thrown together team we held our own and it was a close game. QM3 Josh Calandri was our goalie and became a legend that day making airborne stops that impressed the crowd. All of his hard work still couldn't keep us from losing 3-1. But win or lose the friendship between our two countries was reinforced and we had a great time!

On the first evening we held a reception for about seventy local dignitaries on our flight deck to include the United States Ambassador to Madagascar, Shirley Elizabeth Barnes. It was a nice event and we had the opportunity to meet a few young American women who were there as Peace Corps volunteers.

They were so excited to see Americans! As we talked and got to know each other they had an unexpected though immediately understandable request, they wanted to take a shower and use a toilet. They told me that they had been there for six months and had not taken a hot shower or used a proper toilet since they had arrived. They said they were stationed out in the rural area of town and it was rough. Just from seeing how bad the town was from my walk earlier in the day I couldn't imagine it being any worse. I graciously provided them with my stateroom and told them to take their time. About an hour later they emerged looking very refreshed!

On the 20th Captain Ryan and I were invited to brunch at the Venilla restaurant by the General of the Brigade Raharijaona, Joseph Marie. The United States Ambassador to Madagascar, Shirley Elizabeth Barnes, would also be attending. As the captain and I entered, the general and his staff were there and we exchanged courtesies. There was a language barrier so we did a lot of smiling back and forth.

Our hosts were very gracious and made us feel right at ease. There was a large table set in the sort of open air restaurant. The ambassador arrived, was seated at the head of the table, and we were seated. The food was brought out and it consisted of a plate of mixed cold cuts and what I would call dilled or fermented vegetables. It was an interesting afternoon and you could tell our hosts were out to impress.

*A l'occasion du passage de Monsieur le Ministre des Forces Armées
et de l'escale de l'USCGC SHERMAN*

Le Général de Brigade **RAHARIJAONA Joseph Marie**
Commandant d'Armes de la Place d'ANTSIRANANA

Prie, UNITED STATES COAST GUARD CUTTER " SHERMAN "

*de lui faire l'honneur d'assister au déjeuner (BRUNCH)
qu'il donnera le dimanche 20 mai 2001 à 11 heures
au restaurant VENILLA.*

Tenue correcte/ Petit blanc.

Invitation to the reception

We set sail on the 21st of May and headed for Cape Town, South Africa using the days in transit to conduct drills, helicopter operations, and gunnery exercises. The trip started to get nasty on the 26th of May as we ran into rough gale force 9 seas. As we neared Cape Town on the 26th we received a distress call at 1128 from the 500 foot car carrier *Modern Drive* that was 2 hours away from us to the south at position 33°22.6S 27° 42.3E. She was on fire, listing heavily, and in fear of running aground off of East London, South Africa.

Modern Drive was huge, with a capacity for carrying up to 4500 cars. She was sailing from Freemantle, Australia to Brazil when she went adrift in 30 foot seas with 35 knot winds and an air temperature of about 50 degrees.

M/V Modern Drive

Apparently, several of the 2000 vehicles she was carrying broke loose during the rough weather and crashed spilling gas. The gas ignited and caught the ship on fire. The crew activated the installed firefighting system which extinguished the fire. However, smoke had been sucked into the engine room ventilation system which led the crew to believe they had an engine room fire. Before they realized their mistake, they activated the installed engine room firefighting system which also disabled their main engines.

The first vessel to arrive on scene to render aid was the *M/V Tracer* commanded by Captain Robert M. Baller. She was a heavy lift vessel in route from Abu Dhabi, U.A.E. to Cape Town, South Africa. She had overtaken the *Modern Drive* earlier in the day at around 1030. As she passed the *Modern*

Drive she noticed that the vessel appeared to be floundering and moving violently in the gale force 9 sea state. Captain Baller hailed the *Modern Drive* on the radio and was informed that she was fine and did not require any assistance.

M/V Tracer

At 1130 a distress call was received by the *Tracer*, via radio from East London Control that a car carrier was in distress. It was unclear to *Tracer* what the name of the car carrier was, but its location roughly matched that of the car carrier *Tracer* had passed earlier. Captain Baller made contact via radio with the stricken car carrier, *Modern Drive*, and informed her that they were about 1 hour away and were in route to her location.

As *Tracer* pounded through the gal force 9 seas she maintained radio contact with the *Modern Drive*. A plan to abandon ship was readied for the crew of twenty Filipinos, but put on hold. *Modern Drive* was requesting to be evacuated via helicopter. *Tracer* did not have a helicopter and *Sherman*, who was in radio contact, could launch a helicopter but in the current rough sea state could not recover it.

*M/V **Modern Drive** and M/V **Tracer***

At 1225 *Tracer* arrived on scene of *Modern Drive* and assessed the situation. Captain Baller found *Modern Drive* to be smoking, listing heavily to starboard, and her crew huddled in the wheelhouse. *Sherman* arrived on scene at 1338 and took up a position astern of *Modern Drive*. *Sherman* and *Tracer* held close to *Modern Drive* and assessed the best course of action. At 1445 *Sherman* relieved *Tracer*, who proceeded on her route into Cape Town. Also at 1445 East London Control dispatched the sea tug *M/V Impunzi* to assist.

At 1900 *Sherman's* logs indicate sustained winds peaking at 40 knots.

At 2134 East London Control recalled *M/V Impunzi* due to severe weather conditions. *Sherman* requested to remain on scene with *Modern Drive* until another tug could be dispatched.

Settling in for the night *Sherman* stood by the *Modern Drive* in miserable weather until daybreak on the 27th when we could at least see what we were doing. The *Modern Drive* was about three miles from land and we could see the shore line and buildings of East London. We decided to get a tow line over to her hoping to at least hold her in place until a sea tug could come out and get her. She was about ten times our size, so holding her was going to be about as big of a challenge as getting a tow rope to her. As we took heavy rolls we managed to get the small boat into the water and the boarding team arrived alongside her. They would assess the damage on *Modern Drive* and assist with the tow rope hook up. She was listing pretty heavily, about 20 degrees, to her starboard side and the *Modern Drive* crew had put down a rope ladder for our boarding team to scale.

It was about a good 60-80 foot climb for the boarding team; consisting of LT Ben Strickland, CWO Dave Bellairs, and BM2 Noah Alberici. Watching from *Sherman's* deck I could see them struggling to make progress up the ladder. With the winds blowing them around, wearing cold weather gear, and the weight of their equipment, they would go about 10 or 15 feet and have to stop and rest, just dangling there. Eventually they made it and the next step was getting the huge eight inch thick tow line over to them.

Scaling ladder and our small boat alongside M/V Modern Drive

Sherman maneuvered as close as possible in the rough sea state to pass the tow line messenger, which is a smaller line used to haul the bigger eight inch line over. When we swung around to pass the tow we were only about 50 yards from the *Modern Drive* and she towered over us.

At 1424 we were able to get the tow line set up and keep her stable as we waited for a local sea tug to arrive and take over. At 2245 the sea tug *Pentow Salvor* arrived on scene to relieve us. At 2327 *Sherman* released the tow and turned the *Modern Drive* over to the *Pentow Salvor* who towed her to Algoa Bay for evaluation and then berthed for repairs at Port Elizabeth Harbor. We were now free to resume our transit into Cape Town.

CWO Bellairs, BM2 Alberici, *Modern Drive* crewmember, and LT Strickland aboard *M/V Modern Drive*.

Hooking up the tow rope to the *M/V Modern Drive*

YNC Mark Planitz keeping an eye on the *M/V Modern Drive* tow

Burnt vehicles off-loaded from *M/V Modern Drive*

On the 29th, a day later than originally scheduled, we pulled in starboard side to the No. 2 jetty at the V&A Waterfront in Cape Town, South Africa. It was an overcast and rainy day with low cloud cover but you could still make out Table Top Mountain, a Cape Town landmark.

The port call schedule for Cape Town was as follows;

VISIT
BY
USCG SHERMAN

TO CAPE TOWN
THE REPUBLIC OF SOUTH AFRICA
29 MAY - 02 JUNE 2001

29 May

0730   Admiral Louw (South African Navy) departs Ysterplaat airbase by helicopter to meet Sherman.

0915   Sherman moored #2 jetty at V&A Waterfront

1000   Admiral Louw (South African Navy) departs Sherman

1015   Courtesy call Mayor of Cape Town

1015   Courtesy call Senior Magistrate

1100   Courtesy call Brigadier General Mangethe, Admiral Louw, and Captain Green

1230   Lunch

1300   Rehearsal for Legion of Merit Medal Ceremony

1500   Press Conference

1530   Courtesy call Port Captain Town Harbour and V&A Port Captain

1650 – 1655   Arrival of SA Navy guests for reception and Legion of Merit Medal Ceremony

1700 – 1900 Reception and Legion of Merit Medal Ceremony.

30 May

1000   Wine tour along the Boland/Stellenbosch wine route

1600   Wine tour returns to Sherman

31 May

1300   Sherman's crew departs for Simon's Town Dock Yard

1400   Crew arrives at Simon's Town and greeted by Commander Bisset

1400 – 1545   Tour of Simon's Town dock yard and Museum

1550   Depart for RPC (the club on the West Yard)

1600   Arrive at RPC

1800   Sherman crew depart

1900   Sherman crew arrives #2 jetty

01 June

0815   Sherman Law Enforcement Training Team greeted by Commander Mare (SA) at the Maritime Warfare Training Center in Simon's Town

0830   Sherman volleyball team departs for Wynberg Indoor Sports Complex

0830 – 1030   Law Enforcement at Sea Seminar

1000 – 1315   Volleyball

1315   Lunch

1430   Sherman crew departs for ship

1515   Sherman crew arrives at ship

02 June

TBD   Shift Sherman's berth to berth A for refueling and departure

USCGC Sherman at "full dress" in Cape Town, South Africa

Moored just ahead of us was the South African Navy River Class Mine Hunter *SAS Umhloti M1212*. Her engineering crew had come over to take a tour of our engine room during the port call. After the tour we were invited over to their vessel for a tour. In true military custom, we traded insignia. They were also very impressed with our utility tools that we wore on our belts. I gave mine to one of the engineers as a good will gesture.

Cape Town was a great port call and probably my favorite stop on the trip. One of the highlights for me was taking a tour of the Bergkelder Winery, a famous South African winery located in the historic town of Stellenbosch. Another highlight was visiting with the local South African Navy who invited *Sherman's* crew to their base at Simon's Town, just south of

Cape Town, for a tour. Of course after the tour we ended up at the various junior and senior enlisted clubs for drinks and plenty of sea stories!

South African senior enlisted with HSC Liz Beck and me at Simon's Town Naval Base, South Africa

The chiefs and I ended up at the senior enlisted club. As we settled in we began to swap uniform insignia which was a big hit with our hosts. As we swapped sea stories I mentioned our promotion ceremony to chief petty officer, which was their equivalent to senior enlisted. They said they had a tradition of having their newly promoted member drink a series of alcoholic beverages, one after the other without stopping, as an initiation or rite of passage. If I remember correctly I think it was called a "springbok." It consisted of three glasses of a green colored local liquor, a glass of local wine, and a can of local beer.

It just so happened that we had several chiefs who were preparing for their promotion. When I mentioned this to our hosts they immediately had the bar set up with their initiation drink for them. After our chiefs proved their salt by drinking the combination of drinks they were congratulated by our hosts and welcomed into the South African senior enlisted fraternity!

The traditional South African Navy initiation

Along with the usual formalities of greeting and hosting dignitaries the crew was involved in law enforcement seminars with the South African Navy and we played them in volleyball. I don't know what the score was but the captain wrote home to the ombudsman "don't ask, it was bad!"

Our community outreach project was to participate in Project Crysolice, a boot camp style program that provides at-risk youth the opportunity to learn valuable job skills. *Sherman* volunteers met with the youth and discussed the missions of the U.S. Coast Guard, personal experiences, and American culture. Afterwards the youth were taken on a tour of *Sherman*.

On the 2nd of June *Sherman* was once again underway at 1515. The transit was consumed with daily helicopter operations, drills and non-judicial punishment proceedings. Yes, even Coast Guardsmen get in trouble! Throughout the deployment Captain Ryan took the available time to rule over infractions of the Uniform Code of Military Justice (UCMJ). Punishments ranged from probation to discharge from the service. As the command enlisted advisor I was present at all the proceedings. Although unpleasant, maintaining discipline is a very heavy responsibility and one the captain took seriously.

After Cape Town we headed up the west coast of Africa and crossed the equator at the Prime Meridian on the 9th of June. This is the point on the globe where latitude is zero (0) and longitude is zero (0). King Neptune granted us the status of Emerald Shellback, the rarest of the Shellbacks because of its odd location. It's tucked up on the west coast of Africa near the countries of Nigeria and Ghana. *Sherman* marked the occasion with a swim call.

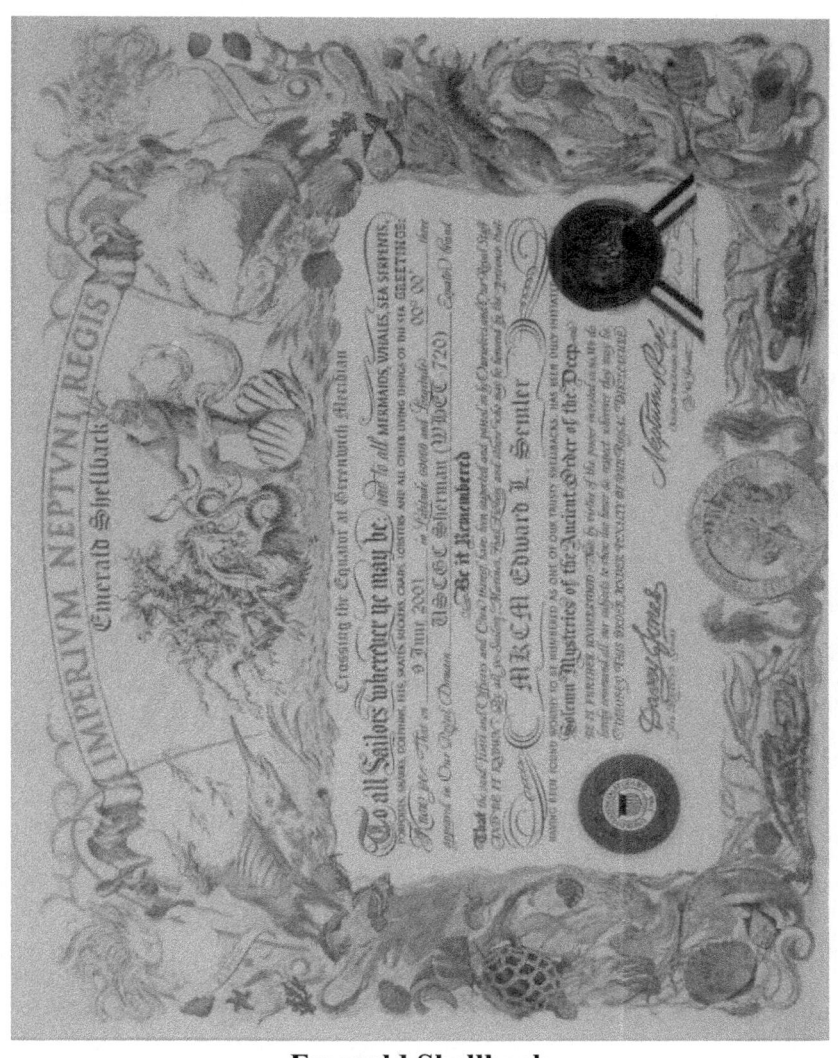

Emerald Shellback

From the 14th to the 16th of June we were in the country of Cape Verde. We moored port side to berth 2 at the island of St. Vicente, or Sao Vicente, in the Porto Grande within the city of Mindelo. The islands which make up Cape Verde are just off the west coast of Africa. The entrance to the city of Mindelo was surrounded by giant mountain ranges and the bay had these huge rocks jetting out of the water. It was spectacular scenery.

View of the Porto Grande with the city of Mindelo in the background

As usual we held a formal reception for the local dignitary's on the cutter's flight deck. And honoring the custom at the Porto Grande, we documented our stop by painting a picture of the cutter on the sea wall. The sea wall was lined with these paintings and it was impressive to see as you walked down the pier into town.

Crewmembers also volunteered their time to paint a local elementary school inside and out.

Back home in Alameda family members were starting to get the "homecoming bug" and held a banner decorating party at the appropriately named Banner club. The event was filled with painting banners, eating, and having a great time!

***Sherman* marking its port call in Cape Verde**

Then it was across the Atlantic Ocean. *Sherman* used the transit time to conduct engineering drills, hold fish call, and of course helicopter operations. Even though we were transiting for home, as with all deployments, we were always on the lookout for illegal activity and available for search and rescue.

Next stop was Bridgetown Barbados, West Indies from the 22nd to the 26th of June. This was another port call that brought many family members and friends out from Alameda. Love was in the air and Captain Ryan reported to the ombudsman that several engagements were announced!

Our last official port call was in Oranjestad Aruba, Dutch Antilles from the 28th to the 30th of June. It's a major tourist destination for Americans so we blended right in with the rest of the American tourist and our arrival was unnoticed. But for us being around Americans meant we were getting closer to home!

As we left Aruba we were still vigilant for illegal activity and moved to intercept the *M/V Manila Pride 6* who was cruising along at 15 knots. After hailing her on VHF channel 16 we caught up with her and after further questioning she was released.

Next it was on to transiting the Panama Canal. We arrived at the entrance to the canal at 0600 on the morning of the 2nd of July. We were directed to an anchorage and set anchor to wait our turn. While at anchorage we conducted Coast Guard business and boarded the M/V Albacore II and M/V Jebel. At 1641 we weighed anchor and were directed to enter the first lock. The Panama Canal is a series of locks and as we were transiting from the Caribbean Sea to the Pacific Ocean we needed to go up in elevation, and this was accomplished via the locks. Although it would take us about 20 hours to transit the canal, the alternative was to sail all the way around the southern tip of South America, which would add weeks to our trip.

We were towed through the locks by locomotive engines on a rail system. We hooked up to locomotive PCC #57 and started our journey through the first set of locks known as the Gatun Locks. At 0110 the morning of the 3rd of July *Sherman* sailed out of the Miraflores Locks, the final locks on the Panama Canal, and into the Pacific Ocean. In transiting the Panama Canal we earned the "Order of the Ditch" certificate, another nautical milestone.

Order of the Ditch

Entering the Panama Canal

Transiting the Panama Canal

As *Sherman* steamed off the west coast of Central America on the 3rd of July the lookout spotted black smoke on the horizon. We quickly moved in that direction to investigate. When we arrived on scene we found the Costa Rican *M/V Ingrid* adrift. They had lit blankets on fire to get our attention. She had been adrift for 10 days after losing power and had been at sea a total of 20 days. Her crew of five had been surviving on fish and what little water they had brought with them.

Costa Rican *M/V Ingrid* adrift

She was a small 70 foot white and blue fishing vessel and was weighed down with nets and fishing buoys. We provided the crew with food, water, and took her in tow headed for Costa Rica. Ironically, the Costa Rican Coast Guard came out in their patrol boat to retrieve the tow on the 4th of July.

We were surprised when we recognized they were operating an old U.S. Coast Guard 82 foot patrol boat we had given them.

The patrol boat was formerly the *USCGC Point Camden (WPB-82373)*. Commissioned in 1970 she had served the U.S. Coast Guard until decommissioned in 1999 and given to the Costa Rican's. She was in great shape with a new paint scheme and they had renamed her *Santa Maria (82-2)*.

With the tow completed it was on to one more stop in San Diego.

Costa Rican Coast Guard cutter *Santa Maria (82-2)*

On the 6th of July *Sherman* was once again called upon for a search and rescue case. The *M/V Furthers'* emergency position-indicating radio beacon (EPIRB) had gone off and *Sherman* along with a Navy P-3 aircraft, call sign *"Newport 06"* were diverted to investigate. When we arrived on seen of the EPIRB's last position we launched CG-6596 to conduct an aerial search. After several hours the *Further's* EPIRB signaled that it was in Colon, Panama and the search was cancelled. *Sherman* was released to proceed home.

While steaming 50 nautical miles south of Cabo San Lucas on the 9th of July we conducted a passenger transfer with the *USS Scout (MCM-8)*. Apparently one of their crewmembers had medical problems and needed to go back to San Diego. He had threatened to kill himself and assaulted a corpsman. After embarking him on *Sherman* he was noted as being calm and presented no threat. Flight Quarters was set and he was flown off the cutter.

On the 11th of July we pulled into the Navy's Fleet and Industrial Supply Center (FISC) Pier in San Diego, California. This was our first United States port call and we needed to clear U.S. Customs and Agriculture.

It would be a short and busy port call. While in San Diego we would embark family members participating in a tiger cruise. A tiger cruise is sort of like "bring your child to work day" and in this case children were allowed to come aboard and transit to Alameda with the cutter. The kids slept in available berthing spaces, ate meals in the messdeck, and stood watch with their crewmember parent. It was a great way to let our kids see what we had been doing the past six months and experience shipboard life. Who knows, they may grow up to be the next generation of coastguardsmen!

The plan of the day in San Diego was as follows;

0900   Sherman to dock at FISC pier, San Diego

0930 – 1000   Sherman "Tigers" escorted aboard ship

1300   Sherman to depart San Diego and begin transit to Alameda

TBD   Activities planned after leaving the FISC pier and underway include anchoring, flight quarters to detach the helicopter, and UNREPing at sea.

After departing San Diego the helicopter was disembarked and flown ahead of us to Alameda, landing on the *USCGC Morgenthau (WHEC-722)*. From there she would transit back to her home in Kodiak, Alaska. Next we lined up on the *USNS Guadalupe (T-AO-200)* for one last refueling taking on 103,347 gallons of fuel.

The plan of the day for arriving in Alameda on the 13th of July was as follows;

0830   *Sherman* to transit under the Golden Gate Bridge. District 11 personnel and a group called "Friends of the Coast Guard" will sponsor a flower drop in which flowers are to be dropped on the *Sherman* as she passes below.

0830 – 1000   Refreshments provided by the Novato Navy League at the Coast Guard base in Alameda.

0900 – 1000   The USS Hornet Band plays as Sherman transits down the estuary to her mooring.

1000   Sherman docks. After Sherman docks and the crew greets family and friends there will be a brief awards presentation. All crew that have been granted liberty should be allowed to leave by early afternoon.

As we passed under the Golden Gate Bridge that foggy morning I stood out on the starboard weather deck by the small boat. Coast Guard District 11 and "friends of the Coast Guard" had planned a flower drop as we passed under and it was a very moving moment as the flowers rained down on us. At that very moment my future wife was standing on the bridge raining flowers down on us and described the experience as being beautiful and exciting!

Sherman **passing under the Golden Gate Bridge**

I knew that we had just made Coast Guard history and that our journey was coming to an end. I could actually feel the weight of the moment as we glided into San Francisco Bay, past the skyline of San Francisco and Alcatraz Island, and into the estuary. We arrived in Alameda to a warm welcome of friends and family waiting patiently on the pier. The *USS Hornet* band was playing and excitement was in the air. It was over. *Sherman* had just travel 38,000 miles in 181 days.

***Sherman* returning home to the Coast Guard base in Alameda, California**

Vice Admiral Riutta stated upon our return that "Your deployment has been widely recognized as the most successful assignment of a Coast Guard cutter to the Arabian Gulf, conducting Maritime Interception Operations in support of the enforcement of United Nations sanctions. Your success began with the ships superb performance during the pre-deployment workup exercises and carried through the entire six-month deployment."

In recognition of her outstanding performance of duty *Sherman* and her crew were awarded the Armed Forces Expeditionary Medal and the Coast Guard Unit Commendation Award with Operational Distinguishing Device.

But In my opinion the most cherished award came from King Neptune. For circumnavigating the globe he bestowed on us his highest award, the Order of Magellan. There was no initiation because there was no one aboard who held the designation to conduct the ceremony it was so rare.

Circumnavigating the globe is rare in the Coast Guard. The first Coast Guard cutter to accomplish the feat was The *USCGC Eastwind (WAGB-279)* in 1960-61. The next was the *USCGC Southwind (WAGB-280)* back in 1968-69. But the *Eastwind* and *Southwind* were both icebreakers. *Sherman* is the only non-ice breaking Coast Guard vessel to accomplish a circumnavigation

Order of Magellan

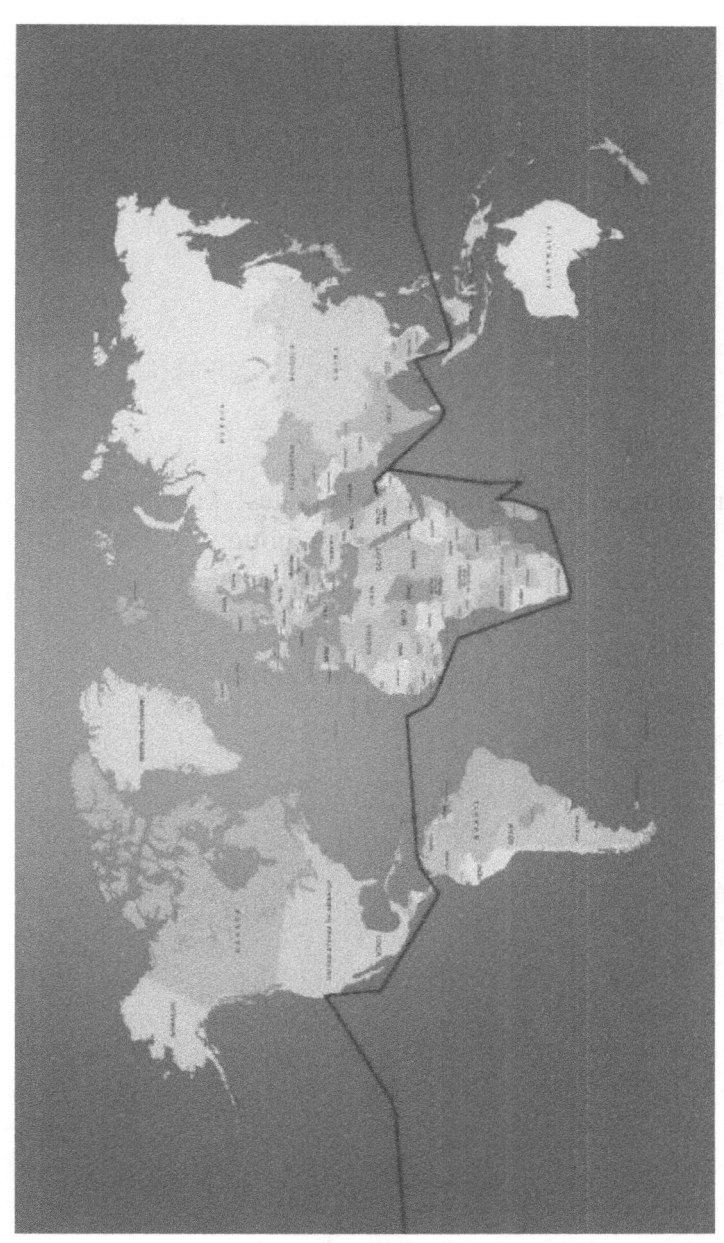

Sherman's circumnavigation route

Fair winds and following seas shipmates! And I will leave you with *Sherman's* motto;

"Honorable and Faithful"

USCGC *Sherman* (WHEC-720) Circumnavigation Crew

Commanding Officer

Captain David Ryan

Executive Officer

Commander David Klipp

Command Enlisted Advisor

Master Chief Machinery Technician Edward Semler

Operations Department

Lieutenant Christopher Randolph
Lieutenant Junior Grade Clayton Beal
Lieutenant Junior Grade Lori Archer
Ensign Robert Rimer
Ensign Theresa Grano
Ensign Yamasheka Young
QMC Robert Eagleton
ETC David Williams
TCC Pamela Arnold
RDC Dennis Ferguson
ET1 Gregg Becky
ET1 Charles Rowell
TC1 Justina Dunn
QM1 Brian Megee
RD1 John Leary

ET2 Anthony Goslin
ET2 Frank Panza
ET2 Philip Ilarde
TT2 Kirk Beckmann
TC2 Christopher Guinther
TC2 Brian Robson
TC2 Greg Gilmore
TC2 Kara Park
TC3 Ethan Corsbie
QM2 Joshua Calandri
QM2 Taran Groom
RD2 Michael Thompson
RD2 Brett Gary
RD2 Ryan Warnke
RD2 Gabriel Flesher
ET3 Jason Fischer
TC2 Devonn Porter
QM3 Kristen Freed
QM3 Jeremy Ballard
RD3 Naeemah Chandler
RD3 Steven Oakford
RD3 Rodolfo Fonseca
RD3 Edgar Mojica
RD3 Jason Rice
RD3 Brenda Hildreth
RD3 Buck

Weapons Department

Lieutenant Junior Grade Benjamin Strickland
Ensign David Yadrick
CWO3 Steven Phillips
FTC John Spotts
GMC Mark Riley
GMC Scott Mercurio
BMC Phillip Spurling
BM1 Steven Achely
FT1 Scott Szcesniak
FT1 Kyle Hallingstad
GM1 Swift
BM2 Mark Evans
BM2 Foy Melendy
BM2 Noah Alberici
GM2 Chad Bowie
BM3 John Percival
BM3 Braden Todaro
BM3 Paul Hendershot
GM3 Daniel Evans
GM3 Andrew Kyle
FT3 Chad Penar
FT3 Michael Zayas
SNBM Marco Cowley
SNBM Brandon Casey
SN Chad Walker
SN Heidi Brady
SN Theresa McElhone
SN Troy Shepard
SN Timothy Playter

SN Ryan Isam
SN Paul Vancore
SN Laura Tonner
SN David Burke
SN Jason James
SN Carolyn Crow
SN Josh Decker
SA Juston Greene
SA Jeremy Dowdy
SA Dale Reed
SA Thadeus Harrah
SA Shannon Kern
SA Michael Starr
SA Robert Potts
SN Mark Henricksen
SN Travis D Hutton
SN Sean Bowles
SN Joe Griffin
SN Sean White
SN Dave McCalister
SN Jim Fitzpatrick
SA Kyle Keene
SN Josh Hart
SN Robert Christy
SN Robert Riggle

Engineering Department

Lieutenant David Socci
Lieutenant Junior Grade Heather Mattern
Ensign Matt Derrenbacher
Ensign Caryn Santogotta
CWO Dave Bellairs
EMCS Joe Barthelemy
EMCS Bob Sumpter
MKC John Young
MKC Dwayne Fontenot
DCC Michael Dewitt
MK1 Patrick Dudley
MK1 Michael Clark
MK1 Andrew Vandewarker
EM1 Thomas Bremer
MK2 Joseph Prince
MK2 James Stetson
MK2 Justin Thompson
DC2 Laura Hatley
MK3 Alan Dowdall
MK3 Sean Sullivan
MK3 Mathew Bolles
MK3 James Sonnier
MK3 Hauck
DC3 Jesus Suarez
DC3 Jeremey Haas
EM3 Lyle Bowen
EM3 Lavelas Luckey
EM3 Jared Tester
SN Joseph Green

FN Mark Stockholm
FN Jon Hammermeister
FN Kort Huettinger
FN Havier Mendoza
FN Tremayne Hicks
FN Brett Newman
FA Jared Wilkenson
FA James Grier
FA Jared Deffries
FA Corey Sisson
SN Stockholm
FN Tony Harris

Supply Department

CWO Paul Sepp
CWO2 Ken Murrell
HSC Elizabeth Beck
FSC Tina Sondrini
YNC Mark Planitz
FS2 Traci Willoughby
FS2 Shayla Barfield
SK2 Dwight Wilson
FS3 Anthony Bullard
FS3 Terry Beshears
SK2 Christopher Stacey
SK3 Kenneth Hailey
SK3 Leisha Stockton
YN3 Melanie Byrd
HS2 Mark A. Gray

SAFS Jason Trescott
FS3 Andy K. Pannelli
FS3 Charles Wilson

Aviation Detachment (AVDET) from Kodiak, Alsaka

Lieutenant Commander John Bevilacqua
Lieutenant Dave Billburg
Lieutenant Dan Baravik
AMT1 John Kent
AMT2 Keith Kranenburg
AMT2 Chris Kluyber
AVT2 Russell Merrick
AST3 Dennis Moyer

USCGC *Sherman* (WHEC-720)

Persian Gulf & Circumnavigation Deployment

13 JAN – 13 July 2001

Ship Log highlights for each day

Note – names were redacted in logs received from Coast Guard Headquarters. Only rates were readable. Names were added by author if known.

13 Jan – Moored portside to Berth One Coast Guard Island Alameda, CA. with standard mooring lines doubled under the OPCON and ADCON of COMPACAREA Alameda CA. Ships status is Charlie. All deck, anchor and aircraft warning lights are energized and burning brightly. Material Condition Yoke is set throughout the ship. Receiving potable water, sewage, electricity, and telephone services via shore tie. All hands are on authorized liberty, with the exception of duty section one. Other ships present are USCGC BOUTWELL (WHEC-719). SHERMAN is SOPA. 0920 Set the Special Sea Detail 0949 Underway the Oakland Estuary enroute Pacific Ocean. 1104 Passed beneath Oakland Bay Bridge. 1138 Passed beneath Golden Gate Bridge.

14 Jan – Underway in the Pacific Ocean in position 36° 35.6N 126° 18.1W enroute Pearl Harbor, Hawaii. Both MDE's are on line turning for 15 knots. SHERMAN is under the OPCON, ADCON, and TACON of COMTHIRDFLT, San Diego, CA.

15 Jan – Underway in the Pacific Ocean in position 34° 12.6N 133° 11.9W enroute Pearl Harbor Hawaii. Both MDE's are on line turning for 15 knots 1530 Commenced Gunnery Exercise. 1533 Secured from Gunnery Exercise after expanding 5 rounds of 25mm.

16 Jan – Underway North Pacific Ocean in position 31° 36.3N 139° 40.6W enroute Pearl Harbor, Hawaii. Both MDE's are on line turning for 15 knots 0100 Retarded clocks 1 hr to confirm with +10 Whiskey time now 0000. 1306 Commenced GQI. 1409 Secured from GQI.

17 Jan - Underway North Pacific Ocean in position 28° 33.7N 145° 40.8W enroute Pearl Harbor, Hawaii. Both MDE's are on line turning for 15 knots 0820 Set Flight Quarters Condition I. 0900 CGNR- 6596 off the deck and away to stbd with 03 POB enroute USS Paul F. Foster. 0915 CGNR-6596 on deck USS Paul F. Foster. 0931 CGNR-6596 off deck USS Paul F. Foster enroute Sherman. 0955 CGNR-6596 on deck with primary tie downs in place. 1000 Secured from Flight Quarters Condition I set refueling detail. 1006 Secured the Helo Refueling Detail having transferred 121 gallons of JP-5 to CGNR-6596. 1056 Traversed CG-6596 into hanger with primary and secondary and heavy weather tie downs in place. 1327 Commenced BECCE Drills. 1353 Commenced Main Space Fire Drill. 1411 secured from BECCE and Main Space Fire Drills. 1457 Commenced CIWS Gunnery Exercise. 1504 Secured from Gunnery exercise having expanded 300 rounds of 20mm with no apparent casualties.

18 Jan - Underway North Pacific Ocean in position 25° 47.4N 150° 20.3W enroute Pearl Harbor, Hawaii. Both MDE's are online turning for 16.5 knots <u>1325</u> Set the Underway Replenishment Bill. <u>1349</u> Commenced approach on the USNS Yukon (T-AO-202). <u>1354</u> Alongside the USNS Yukon for personnel and cargo transfer and refueling. <u>1421</u> Commenced refueling. <u>1430</u> Commenced preparations for entering port in Pearl Harbor, Hawaii. <u>1459</u> Completed fueling having received 35641 gallons of diesel fuel. <u>1501</u> Commenced Emergency Breakaway Drill. <u>1516</u> Secured the Underway Replenishment Bill.

19 Jan - Underway North Pacific Ocean in position 22° 03.6N 155° 58.6W enroute Pearl Harbor, Hawaii. NR2 MGT is online turning for 20 knots. <u>0700</u> Set the Navigation and Anchor Detail. <u>0817</u> Moored starboard side to Pier B26 Pearl Harbor, Hawaii. <u>0820</u> Secured from Special Sea Detail. <u>1415</u> Liberty granted to all hands with the exception of duty section 2.

20 Jan – Moored starboard side to Pier B26 Pearl Harbor, Hawaii. Other ships present are USS Lake Erie (CG-70), USS Port Royal (CG-73). USS Lake Erie is SOPA. <u>0845</u> Liberty expired for duty section 3. Held morning muster for duty section 3. <u>0905</u> Liberty granted to duty section 2.

21 Jan - Moored starboard side to Pier B26 Pearl Harbor, Hawaii. Other ships present are USS Lake Erie (CG-70), USS Port Royal (CG-73). USS Lake Erie is SOPA. <u>0847</u> Set the Special Sea Detail. <u>0941</u> Secured Special Sea Detail with the exception of Navigation Detail. <u>1011</u> Crossed Line of Demarcation switched to International Navigation Rules. <u>1030</u> Traversed CG-6596 from flight deck to the hanger, primary, secondary, and heavy weather tie downs in place.

22 Jan - Underway North Pacific Ocean in position 21° 19.6N 162° 26.4W enroute Apra Harbor, Guam. NR1 MGT is online turning for 21 knots. 0836 SHE-2 lowered to the rail. 0820 Late entry traversed CGNR-6596 to the flight deck, primary and secondary tie downs in place. 0847 SHE-2 lowered to the water with BM2 as coxswain, BM2, BM3, and MK3 as boat crew. 0849 SHE-2 away to port for small boat training. 0853 Set Flight Quarters Condition I. 0819 Helo off the deck and away to port with 04 POB. 0830 Secured from Flight Quarters Condition I, set Flight Quarters Condition 2. 1019 Secured from Flight Quarters Condition 2 set flight Quarters Condition 1 for landing of CGNR-6596 with 04 POB. 1052 CGNR-6596 on deck, commenced touch and go landing. 1107 CGNR-6596 over deck after completing 04 touch and go landings, commenced Hot Refuel. 1111 Secured Hot Refuel having transferred 131 Gallons of JP-5 to Helo. 1112 CGNR-6596 on deck with primary and secondary tie downs. 1120 Secured from Flight Quarters. 1822 Set the Boarding Detail, Boarding Team consists of RD2, SN, GM3, FT2, EN, SN, SN, FN Kort Huettinger, BM3, BM2, SN, MK3, SN, FN, SN, and LTJG is the boarding officer. 1838 SHE-2 lowered to the water with BM2 as coxswain, SN and MK3 and 08 boarding team members the boat crew. 1842 SHE-2 Underway enroute M/V Yukon. 1854 SHE-2 alongside M/V Yukon. 1859 Boarding team safely on board M/V Yukon. 1903 SHE-2 alongside to port. 1907 SHE-2 Underway with 11 POB enroute M/V Yukon. 1910 Remaining barding team members safely on M/V Yukon. 1914 SHE-2 alongside to port. 1918 SHE-2 raised to the main deck rail with coxswain and boat crew safely aboard SHERMAN. 2124 SHE-2 lowered to the water with BM2 as coxswain and MK3 boat crew, completed boarding of M/V Yukon. 2132 08 members of the boarding departed M/V Yukon safely on SHE-2 and Underway enroute SHERMAN. 2136 SHE-2 alongside to port. 2138 08 passengers safely aboard SHERMAN. SHE-2 Underway enroute M/V Yukon. 2147 08 members of boarding team departed M/V Yukon safely on SHE-2 and Underway enroute SHERMAN 2151 SHE-2 alongside to port crew aboard. 2200 Released USNS Yukon 2400 Retarded clocks 1 HR to conform to +12 YANKEE time now 2300.

23 Jan - Underway North Pacific Ocean in position 21° 06.5N 169° 0.1W enroute Apra Harbor, Guam. NR2 MGT is online turning for 19 knots. 0834 Set Flight Quarters Condition I. 0902 Helo off deck and away to sbd with 04 POB enroute USS STETHAM. 0905 Secured from Flight Quarters Condition I. 1029 Set Flight Quarters Condition I for landing CGNR-6596 enroute SHERMAN with 04 POB. 1042 CGNR-6596 on deck with primary and secondary tie downs. 1043 Commenced Hot Refuel with Helo. 1048 Secured Hot Refuel on deck having transferred 127 gallons of JP-5 to Helo 1052 Helo secured engines. 1053 Secured from Flight Quarters and secured #1 MDE due to casualty.

24 Jan - Underway North Pacific Ocean in position 20° 35.9N 176° 42.0W enroute Apra Harbor, Guam. NR1 MGT is online turning for 21 knots. 0720 Set the Underway Replenishment Bill. 0804 Commenced approach on the USNS Yukon. 0811 Alongside USNS Yukon for refueling. 0839 Commenced refueling 0913 crossed 180 Meridian at 20 19.7 N Latitude. 0947 Completed fueling having transferred 89,273 gallons of JP-5. 0957 Secured the UNREP Bill. 1012 Set Flight Quarters Condition I for launch. 1046 CGNR-6596 off deck an away to starboard, 04 POB. 1102 CGNR-6596 on deck STETHEM.

1121 CGNR-6596 off deck USS STETHEM enroute SHERMAN. 1127 CGNR -6596 on deck primary and secondary tie downs in place. 1200 Advanced clocks 24 hours to conform with -12 MIKE time zone. Time now 1200 25 January 2001, 1204 Secured Flight Quarters. 1301 Set the Helo Refueling Bill. Secured the refueling detail having transferred 90,273 gallons of JP-5 fuel to CGNR-6596 (obviously a bad log entry)

26 Jan - Underway North Pacific Ocean in position 19° 54.5N 176° 29.2E enroute Apra Harbor, Guam. Both MDE's are online turning for 16.5 knots. SHERMAN is under the OPCON, ADCON, and TACON of COMUSNAVCENT SEVENTHFLT. 0100 retarded clocks one hour to conform with -11 LIMA time now 0000. 1328 Commenced General Quarters Drill. 1414 Secured from General Quarters Drill. 1417 Visibility reduced to 4,000 YDS, engaged all standard navigation lights. 1429 Visibility increased to 12,000 YDS secured all standard navigation lights

27 Jan - Underway North Pacific Ocean in position 19° 06.7N 170° 15.0E enroute Apra Harbor, Guam. Both MDE's are online turning for 16.5 knots. 1335 Traversed CGNR-6596 onto flight deck with primary and secondary tie downs. 1426 Traversed Helo into hanger with primary, secondary, and foul weather tie downs. 1458 Energized standard navigation lights due to reduced visibility. 1508 Secured all standard navigation lights due to increased visibility.

28 Jan - Underway North Pacific Ocean in position 18° 24.9N 164° 59.3E enroute Apra Harbor, Guam. Both MDE's are online turning for 15.5 knots.

29 Jan - Underway North Pacific Ocean in position 17° 32.6N 158° 47.9E enroute Apra Harbor, Guam. Both MDE's are online turning for 16.5 knots. 1418 Traversed CGNR-6596 onto flight deck, primary and secondary tie downs in place. 1445 Set Flight Quarters Condition I for launch. 1521 Helo off the deck and away to port with 03 POB. 1620 Helo completed 08 touch and go landings and 02 landings with Talon engagement, enroute USS STETHEM with 03 POB. 1637 CGNR-6596 on deck USS STETHEM. 1643 CGNR-6596 away from USS STETHEM with 05 POB enroute SHERMAN. 1654 CGNR-6596 on deck, primary tie downs in place. 1655 Disembarked 02 POB. 1656 Commenced Hot Refuel on deck with Helo. 1659 Secured Hot Refuel on deck having transferred 103 gallons of JP-5 to Helo. 1705 Secured Flight Quarters. 1757 Set Flight Quarters Condition I. 1827 Helo is off deck and away to port with 03 POB. 1854 Helo on deck after completing 03 touch and go landings. 1855 Commenced Hot Refueling on deck with Helo. 1859 Secured Hot Refueling after transferring 71 gallons of JP-5 to CGNR-6596. 1902 Secured Flight Quarters Condition I.

30 Jan - Underway North Pacific Ocean in position 16° 02.4N 152° 51.0E enroute Apra Harbor, Guam. NR2 MGT is online turning for 22 knots. 1232 Traversed CGNR-6596 onto flight deck, primary and secondary tie downs in place. 1250 Set Flight Quarters Condition I. 1314 Helo is off the deck and away to port with 04 POB. 1316 Secured Flight Quarters Condition I set Flight Quarters Condition II. 1328 Commenced General Quarters Drill. 1434 Helo on deck with primary and secondary tie downs. 1440 Secured from General Quarters Drill. 1458 set the Helo Refueling Bill. <u>1506</u> Secured the Helo Refueling Bill after having transferred 96 gallons of JP-5 to the Helo.

31 Jan - Underway North Pacific Ocean in position 14° 07.2N 146° 12.4E enroute Apra Harbor, Guam. Both MDE's are online turning for 16.5 knots. <u>0710</u> Set Flight Quarters Condition I. <u>0735</u> Helo above deck and away port enroute NAVSTA Guam with 05 POB. <u>0747</u> CGNR-6596 on deck NAVSTA Guam. <u>0825</u> Set Navigation and Anchor Detail. <u>0904</u> Moored starboard side to wharf Echo Apra Harbor, Guam. <u>0913</u> Secured Special Sea Detail <u>1005</u> Commenced refueling. <u>1020</u> EM3 Lavelas Luckey was electrocuted while working on Red Gear. Pulse 120 Blood Pressure 140/100, respiration 40 per minute and shallow. <u>1035</u> EM3 enroute emergency room. <u>1305</u> Secured from fueling detail after having transferred 77,019 gallons of JP-5. <u>1315</u> Set the Special Sea Detail. <u>1351</u> Underway Apra Harbor, Guam. <u>1444</u> Moored portside to wharf Victor Apra Harbor, Guam. <u>1448</u> Secured Special Sea Detail. <u>1617</u> Liberty granted to all hands with the exception of duty section 4 to expire 0730 01 Feb 01. <u>2230</u> FN Corey Sisson arrived PCS from TRACEN Cape May, NJ.

01 Feb – Moored portside to berth V3 Apra Harbor, Guam. Other ships present are USCGC GALVESTON ISLAND (WPB-1349) SHERMAN IS SOPA. <u>0833</u> Underway from V3 Apra Harbor, Guam outbound for sea. <u>0911</u> Secured from Special Sea Detail. <u>1450</u> SA Christy fell down ladder into compartment 3-32-0-A (Bosn Stores) transported to sickbay with apparent injuries to mid-section. <u>1550</u> Corpsman consulted with Flight Surgeon on SA injuries. <u>1658</u> Set Flight Quarters Condition I for landing. <u>1735</u> CGNR-6596 on deck with primary tie downs in place <u>1739</u> Secured Flight Quarters

Condition I set the Helo Refueling Bill. 1747 Secured the Helo Refueling Bill having transferred 25 gallons of JP-5 to CGNR-6596 1804 Set Flight Quarters Condition I 1816 Helo is off the deck and away to port with 05 POB enroute Naval Hospital, Guam 1825 Secured from Flight Quarters Condition One set Flight Quarters Condition two 1850 Set Flight Quarters Condition one for landing. 1908 CGNR-6596 is on deck with primary tie downs in place. 1903 Secured from Flight Quarters Condition one set the Helo Refueling Bill. 1920 Secured the Helo Refueling Bill after having transferred 101 gallons of JP-5 to the Helo. 1955 Transferred Helo into hanger with primary, secondary, and foul weather tie downs. 2300 Retarded clocks to conform with -9 India time now 2200.

02 Feb - Underway North Pacific Ocean in position 13° 09.4N 142° 15.9E enroute Singapore. Both MDE's are online turning for 16.5 knots. 0924 SHE-1 lowered to the water with BM2 as coxswain, SN and MK3 as boat crew. SN, SN, SN, SN, SN, FN as passengers. SHE -1 underway for training in position 12° 48,9N, 132° 42.7 E 0930 SHE-1 alongside to starboard. 0931 SHE -1 away to starboard 0940 SHE -1 reporting via VHF channel 83A a coolant leak coming from coolant cap. 0941 SHE -1 D.I.W. 0951 SHE-1 underway enroute SHERMAN. 0958 SHE-1 alongside to starboard 0959 SHE-1 raised to the main deck, coxswain, boat crew and passengers safely aboard. 1348 LONEWOLF 43 enroute SHERMAN from USS PAUL F FOSTER. 1352 Set Flight Quarters Condition 3 for VERTREP 1405 Secured VERTREP having received cargo. 1410 Helo away to port. 1412 Secured Flight Quarters Condition 3. 1438 SHE-2 lowered to the water 1440 SHE-2 away to port with BM2 as coxswain, MK3, SN, SN, SN and FN as boat crew, enroute USS STETHEM. 1518 SHE-2 enroute SHERMAN. 1546 SHE-2 alongside to port 1548 SHE-2 commenced approach drills. 1617 SHE-2 raised to the main deck rail, coxswain and boat crew safely aboard SHERMAN. 1624 SHE-2 cradled and ready for sea.

Note – LONEWOLF was USS PAUL F. FOSTERS SH-60B Helicopter.

03 Feb Underway North Pacific Ocean in position 12° 17.1N 136° 12.2E enroute Singapore. Both MDE's are online turning for 16.5 knots. 1305 SHE-2 lowered to the water with BM as coxswain, SN as boat crew and MK3 as engineer, GM as passenger SHE-2 underway for swim call. 1430 Secured swim call. 1450 SHE-2 raised to the main deck rail coxswain and boat crew safely aboard. 1500 Traversed CGNR-6596 onto flight deck primary and secondary tie downs in place 1535 Set Flight Quarters Condition I for launch 1610 Helo off the deck to port with 05 POB enroute to PAUL FOSTER. 1615 Secured Flight Quarters Condition I 1811 Set Flight Quarters Condition I Helo on deck with 05 POB. 1819 Disembarked 02 POB. 1820 Commenced Hot Refuel on deck with Helo 1822 Secured Hot Refuel on deck having transferred 73 gallons of JP-5 to Helo 1826 Secured Flight Quarters Condition I 1855 CGNR-6596 hangered with primary secondary and foul weather tie downs in place. 2300 Retarded clocks back one hour to conform with -8 Hotel. Time on deck 2200.

Note – Swim call was held over the Marianas Trench

04 Feb - Underway North Pacific Ocean in position 11° 20.3N 130° 30.5E enroute Singapore. Both MDE's are online turning for 17 knots. 1030 Declutched NR1 MDE due to high jacket water temperature. 1056 Declutched NR2 MDE due to loss of start air pressure 1059 Clutched in NR2 MDE 1343 Declutched NR2 MDE, clutched in NR2 MGT. 1725 Commenced transit passage of Philippine Territorial waters.

05 Feb - Underway in the Bohol Sea in position 09° 36.4N 124° 50.8E enroute Singapore. Both MDE's are online turning for 15 knots. 1434 Traversed CGNR-6596 onto flight deck primary and secondary tie downs in place 1505 Set Flight Quarters Condition I for launch. 1543 Helo off deck and away to port with 05 POB, enroute USS STETHEM. 1553 Secured Flight Quarters Condition I set Flight Condition 2. 1623 Set Flight Quarters Condition I for landing 1631 CGNR-6596 on deck, primary tie downs in place, disembarked 02 passengers, embarked 03 passengers. 1636 Removed primary tie downs from Helo. 1537 Helo off the deck and away to port with 05 POB enroute USS STETEHEM. 1645 Set Flight Quarters Condition I for landing 1702 Helo on deck, primary tie downs in place disembarked 02 POB 1704 Helo away to port with 03 POB. 1706 Set Flight Quarters Condition 4 for HIFR. 1710 CGNR-6596 over the deck. 1713 Passed refueling hose to Helo, commenced HIFR. 1719 Secured from HIFR having transferred 125 gallons of JP-5 to Helo. 1720 Helo away to port, secured Flight Quarters Condition 4, set Flight Quarters Condition I for landing. 1724 CGNR-6596 on deck with primary tie downs in place 1824 Traversed CGNR-6596 into hanger primary, secondary, and heavy weather tie downs in place

06 Feb - Underway in the Sulu Sea in position 08° 26.0N 119° 35.8E enroute Singapore. Both MDE's are online turning for 15 knots 0840 Entered Philippine Territorial waters. 1137

Exited Philippine Territorial waters 1205 entered Philippine Territorial waters 1232 Excited Philippine Territorial waters 1610 traversed the Helo to flight deck with primary tie downs. 1730 Set Flight Quarters Condition one for launch. 1817 Helo off deck and away to port with 04 POB. 1826 Helo on deck after completing two landings 1827 Helo off deck and away to port with 04 POB 1856 Helo off deck and away to port after completing 9 touch and go and 2 landings 1903 Set Flight Quarters Condition 4 for HIFR. ???? Passed refueling hose to Helo, commenced HIFR 1920 secured from HIFR 1922 Recovered fueling hose from Helo 1923 Helo away to port. Secured Flight Quarters Condition 4 set Flight Quarters Condition one for landing 1934 Helo on deck with primary tie downs in place 1951 Secured from Flight Quarters Condition one. 2021 traversed Helo into hanger with primary, secondary and foul weather tie downs in place.

07 Feb - Underway in the South China Sea in position 06° 55.4N 116° 08.8E enroute Singapore. Both MDE's are online turning for 15 knots 0815 Commenced small arms familiarization training with 9mm, M-16, and shotgun. 0907 Secured from small arms fire. 0910 Commenced Gunnery Exercise .25 cal. 1342 Traversed Helo with primary and secondary tie downs in place. 1527 Set Flight Quarters Condition I for launch 1556 Helo off the deck and away to port with 03 POB enroute USS PAUL F FOSTER. 1553 Secured Flight Quarters Condition I set Flight Quarters Condition 2. 1619 Set Flight Quarters Condition I for landing, CGNR-6596 enroute from USS PAUL F FOSTER with 03 POB 1630 CGNR-6596 on deck primary tie downs in place. 1632 Helo off deck and away to port with 03 POB 1638 CGNR-6596 on deck with primary tie downs 1641 CGNR-6596 off deck and away to port with 03 POB 1642 CGNR-6596 on deck. 1643 CGNR-6596 off deck and away to port with 03 POB. 1645 CGNR-6596 on deck with primary tie downs in place 1649 Secured Flight Quarters Condition 1, set Flight Quarters Condition 2, set the Helo Refueling Bill. 1656 Secured the Helo Refueling Detail having transferred 0 gallons of JP-5 to CGNR-6596. 1741 Commenced Refueling CGNR-6596 on deck. 1744 secured Refueling Helo on deck after having transferred 57 gallons 1755 Set Flight Quarters Condition 1 1808 CGNR-6596 above deck and away to port with 03 POB to commence touch and go landings. 1819 Secured Flight Quarters Condition one after completing 03 touch and go landings set Flight Quarters Condition four for HIFR 1828 CGNR-6596 over the deck 1831 Passed refueling hose to Helo, commenced HIFR 1833 Secured pump, recovered refueling hose from Helo 1855 Secured from Flight Quarters Condition 4, set Flight Quarters Condition 1 for touch and go landings. 1925 Secured Flight Quarters Condition 1 after completing 09 touch and go landings, 03 with primary tie downs. Commenced Helo Refueling Bill 1932 Secured Helo Refueling Bill after having transferred 67 gallons of JP-5 to Helo

8 Feb – Underway in the South China Sea in position 04° 49.7N 110° 21.2E enroute Singapore. NR1 MGT online turning for 22 knots. 0756 Traversed CGNR-6596on to flight deck, primary and secondary tie downs in place. 0820 Commenced all preparations for entering port in Sewbawang shipyard, Singapore. 0840 Set Fight Quarters Condition I for launch, removed secondary tie downs from Helo. 0915 Helo off the deck and away to port with 03 POB enroute USS

STETHEM. 0926 CGNR-6596 enroute SHERMAN with 05 POB. 0926 Helo on deck, primary tie downs in place, disembarked 02 passengers. 0935 Embarked 02 passengers. 0940 Helo off the deck and away to port with 05 POB, enroute USS STETHEM. 0948 SHE-1 lowered to the water with BM3 as coxswain BM2, MK3 and SN as boat crew. 0950 SHE-1 away to starboard for boat ops. 0955 SHE-1 alongside to starboard. 1002 CGNR- 6596 enroute SHERMAN with 04 POB. 1005 Helo on deck, primary tie downs in place, disembarked 02 passengers. 1008 Helo off the deck and away to port with 03 POB, secured Flight Con 1, set Flight CON 3 for VERTREP 1015 SHE-1 raised to the main deck rail, coxswain and boat crew safely aboard SHERMAN 1019 Commenced VERTREP for training, set Flight Quarters Condition 1 for landing 1043 Helo on deck, primary tie downs in place. 1046 Commenced Hot Refuel on deck with Helo 1049 secured Hot Refuel having transferred 111 gallons of JP-5 to Helo. 1053 Secured Flight Quarter Condition 1. 1059 SHE-1 cradled and ready for sea.

09 Feb - Underway in the South China Sea in position 01° 33.8N 104° 49.8E enroute Singapore. Both MDE's are on line turning for 10 knots. 0450 Commenced preparations for entering port in Singapore Harbor. 0700 Set the Special Sea Detail. 0849 Moored starboard side to Bert S-5, Sembawang Shipyard, Singapore. 0859 Secured the Special Sea Detail. 0932 The OOD shifted the watch from the bridge to the quarterdeck. 1235 QM3 reported aboard TAD.

10 Feb – Moored starboard side to Berth S-5 Sembawang shipyard, Singapore. All hands are on liberty with the exception of duty section 1. Other ships present are USS

STETHEM and USS FOSTER. 0154 SA returned aboard from Singapore Naval Hospital. 0850 Held morning muster for duty section one and two all hands are present or accounted for 0900 Liberty is granted to duty section one.

11 Feb - Moored starboard side to Berth S-5 Sembawang shipyard, Singapore. All hands are on liberty with the exception of duty section two. Other ships present are USS STETHEM and USS FOSTER. 0845 Held morning muster for duty sections two and three, all hands present or accounted for.

12 Feb - Moored starboard side to Berth S-5 Sembawang shipyard, Singapore. All hands are on liberty with the exception of duty section three. Other ships present are USS PAUL FOSTER (DD 964) and USS STETHEM (DDG 63). SHERMAN is SOPA. 0845 Liberty expired for duty section four. Held morning muster for duty sections three and four. All hands present or accounted for.

13 Feb - Moored starboard side to Berth S-5 Sembawang shipyard, Singapore. All hands are on liberty with the exception of duty section three. Other ships present are USS PAUL FOSTER (DD 964) and USS STETHEM (DDG 63). SHERMAN is SOPA. 0845 Liberty expired for duty section one. Held morning muster for duty sections one and four. All hands present or accounted for.

14 Feb - Moored starboard side to Berth S-5 Sembawang shipyard, Singapore. All hands are on liberty with the exception of duty section one. Other ships present are USS PAUL FOSTER (DD 964) and USS STETHEM (DDG 63). SHERMAN is SOPA. 0800 liberty expired for all hands. 0805

Held morning muster. All hands present or accounted for. 1015 liberty granted to all with the exception of duty section 2 to expire 0645 15 Feb 01. 1045 commenced Commanding Officers Non-Judicial Proceedings under Article 112 UCMJ. 1126 Secured from Commanding Officers Non-Judicial Proceedings with the following results, SA is awarded 1 day restriction and is being processed for discharge for violation of Article 112 of the UCMJ. 2130 Ferry boat seen making circles of Paul F. Foster, stops at approx. 20 ft. before patrol boat (text unreadable). Green Hull, White topside. Not anyone on board with exception of captain.

15 Feb - Moored starboard side to Berth S-5 Sembawang shipyard, Singapore. All hands are on liberty with the exception of duty section two. Other ships present are USS PAUL FOSTER (DD 964) and USS STETHEM (DDG 63). SHERMAN is SOPA. 0400 BM2 and SA departed TAD to Alameda Island California. 0550 Shifted from shore power to ships power. 0716 Shifted the watch from the quarterdeck to the bridge, set the Special Sea Detail. 0802 Underway in Johor Straight 0818 Secured the Special Sea Detail with the exception of the Navigation and Anchor Detail. 0958 Pilot vessel alongside to starboard. 0959 Disembarked Singapore Pilot 1000 Pilot vessel away to starboard 1037 Secured the Anchor Detail 1104 Secured the Navigation Detail 1313 Commenced preparations for entering port in Phuket, Thailand. 2300 Retarded ships clocks to conform with -7G time zone, time now is 2200.

16 Feb – Underway in the Strait of Malacca conducting MEFEX-01 in position 03° 50.0N 99° 56.5E enroute Phuket, Thailand. NR1 MGT is online turning for 22 knots. 1230 Set the Navigation and Anchor Detail. 1255 Commenced approach on anchorage Pa tong Bay, Thailand 1256 SHE-2 lowered to the main deck rail with BM2 as coxswain, SN, and MK3 as boat crew. 1303 Anchored in Pa tong Bay, Thailand in position 07° 54.2 N 058° 16.7 E with 3 shots of chain on deck to the starboard anchor to a sand and mud bottom in 12.2 meters of water using the following bearings 056 true to Foxtrot, 112 E to (G) and ranges 3100 yards to "LA" and 2291 yards to "LB" 1317 Secured Navigation and Anchor Detail set the Anchor Watch 1318 SHE-2 lowered to water with 05 passengers 1319 SHE-2 away to port 1540 SHE-2 alongside to port boat crew safely aboard 1628 Boat crew safely onboard SHE-2. SHE-2 away to port enroute PAUL F. FOSTER (DDG-964) 1634 SHE-2 alongside USNS PAUL F. FOSTER. SHE-2 away having transferred 01 passengers. 1718 SHE-2 alongside to port 1720 Liberty is granted to all hands with the exception of duty section 3 to expire on board 0845 17 Feb 01 1902 SHE-2 away to port enroute USS STETHEM and USS PAUL F. FOSTER. 1913 SHE-2 alongside to port.

17 Feb – Anchored in the Pa tong Bay Phuket, Thailand in position 07° 54.2N 098° 16.7E with three shots of chain on deck in 12 meters of water. All hands are on authorized liberty with the exception of duty section three. 0900 liberty granted to duty section 3 to expire 0845 20 Feb 01. 1545 SHE-2 away to port with BM2 as Coxswain and MK3 and SN as boat crew. 1640 SHE-2 alongside to port

18 Feb - Anchored in the Pa tong Bay Phuket, Thailand with three shots of chain on deck in 17 meters of water. Ships status is ALPHA. All hands are on authorized liberty with the exception of duty section four. 0830 Liberty expires for duty section 1 0853 Liberty granted to duty section 4 to expire 1030 22 Feb 01. 0950 SHE-2 underway from port to conduct training with (text unreadable). 1315 SHE-2 away to port for training with SN as coxswain, MK3 as Eng., and BM3 as crew. 1350 SHE-2 alongside to port. 1626 SHE-2 away to port with SN as coxswain, MK3, SN, SA as boat crew. 1641 SHE-2 alongside to port coxswain and boat crew safely aboard SHERMAN.

19 Feb - Anchored in the Pa tong Bay Phuket, Thailand with three shots of chain on deck in 17 meters of water. Ships status is ALPHA. All hands are on authorized liberty with the exception of duty section one. 0530 SHE-2 away to port for training with SN as coxswain, MK3 as ENG, and SN as crew. 0555 SHE-2 alongside to port 0845 Held muster for duty sections 1 and 2. All hands are present or accounted for. 0910 Liberty granted for duty section one to expire 1030 22 Feb 01. 1326 SHE-2 away to port with BM3 as coxswain and MK3 as boat crew. 1520 SHE-2 alongside with coxswain and boat crew safely aboard. 1959 SHE-2 alongside disembarking.

20 Feb - Anchored in the Pa tong Bay Phuket, Thailand with three shots of chain on deck in 38 feet of water. Ships status is ALPHA. NR2 SSDG is providing electrical power. All deck, anchor, and aircraft warning lights are energized and burning brightly. Material condition Yoke is set throughout the ship. SHE-2 is alongside to port. SHE-1 is cradled for sea. CGNR-6596 is on deck with primary and secondary tie downs in place. SHERMAN is under the OPCON ADCON and TACON of COMSEVENTH FLT. All hands are on authorized liberty with the exception of duty section two. 0845 Held muster for both duty sections two and three. 0918 Liberty granted to all hands with the exception of duty section three to expire no later than

1000 Feb 22 2001. <u>1612</u> SHE-2 away to starboard with BM2 as coxswain SN, and MK3 as boat crew. <u>1642</u> SHE-2 alongside to port coxswain and boat crew safely aboard

21 Feb - Anchored in the Pa tong Bay Phuket, Thailand with three shots of chain on deck in 40.1 feet of water. Ships status is ALPHA. All hands are on authorized liberty with the exception of duty section three. <u>0403</u> SHE-2 away to port with BM3 as coxswain, SN and MK3 as boat crew. <u>0419</u> SHE-2 alongside to port coxswain and boat crew safely aboard. <u>0845</u> Held muster for duty sections three and four. All hands present or accounted for. <u>0902</u> Liberty granted to duty section three to expire no later than 1030 22 Feb 01. <u>1233</u> SHE -2 away to port with SN as coxswain, SN and MK3 as crew and LTJG as passenger. <u>1236</u> SHE-2 alongside USS PAUL F. FOSTER, embarked 02 passengers, enroute SHERMAN <u>1246</u> SHE-2 alongside to port crew and passengers safely aboard. <u>1330</u> SHE-2 away to port with SN as coxswain and SN and MK3 as crew. <u>1402</u> SHE-2 alongside to port, boat crew safely aboard SHERMAN. <u>1910</u> SHE-2 away to port with SN as coxswain and MK3 as crew, enroute PAUL F. FOSTER. <u>1815</u> SHE-2 alongside to port, boat crew safely aboard SHERMAN. <u>2115</u> SHE-2 safely away to port with SN as coxswain, BM2 and MK3 crew and 02 passengers. <u>2122</u> SHE-2 alongside PAUL F. FOSTER to port disembarked 01 passenger. <u>2135</u> SHE-2 safely alongside to port with coxswain, crew, and passenger safely aboard. <u>2130</u> SHE-2 raised to the port main deck rail

22 Feb - Anchored in the Pa tong Bay Phuket, Thailand with three shots of chain on deck in 16 meters of water. Ships status is ALPHA. All hands are on authorized liberty with the exception of duty section four. <u>0724</u> SHE-2 away to port with

SN as coxswain and MK3 as crew. 0750 SHE-2 alongside to port, boat crew safely aboard. 0820 commenced preparations for getting underway 0825 SHE-2 lowered to the water's edge. 0830 SHE-2 away to port with SN as coxswain, and MK3 and SN as crew. 0835 liberty expires for all hands. 0847 SHE-2 alongside to port with 06 POB, crew safely embarked SHERMAN. 0900 SHE-2 away to port with 5 POB. 0920 set the Nav and Anchor Detail. 0933 SHE-2 alongside to port, boat crew embarked SHERMAN safely. 0935 SHE-2 raised and secured at the deck rail. 0936 Clutched in both MDE's and placed in Pilot House Control. 0940 Commenced heaving around on the port anchor. 0947 anchors aweigh, underway in Pa tong Bay enroute Open Ocean observing International Navigation Rules. 0954 Placed both MDE's in ERC. 0958 hawsed the port anchor. 1000 Secured the Anchor Detail 1006 Secured Navigation Detail, set the sea watch. 1505 traversed the Helo into hanger primary and secondary tie downs. 2300 Retarded ship's clocks 1 hour to conform with -6F time zone.

23 Feb – Underway in the Indian Ocean conducting MEFX-1 patrol in position 06° 24.6N 095° 53.0E. Both MDE's are online turning for 15 knots. 0335 Declutched NR1 MDE due to casualty, ETR unknown. 0739 Commenced procedures for steering casualty. 0745 secured from steering casualty. 0940 Commenced 50 cal. Gunnery Exercise. 1019 secured from 50 cal. Gunnery exercise. 1330 set Flight Quarters Condition 1 for launch removed secondary tie downs from CGNR-6596. 1355 Helo off the deck and away to port with 04 POB 1403 SHE-2 lowered to the water, with BM2 as coxswain SN as boat crew, MK3 as engineer SN and SN as passengers. SHE-2 underway to do training with CGNR-6596. 1507 Set Flight Quarters

Condition I for landing. 1520 CGNR-6596 on deck, primary tie downs in place 1521 Commenced Hot Refuel on deck with Helo. 1525 Secured Hot Refuel on deck having transferred 94 gallons of JP-5 to Helo 1528 Helo off the deck and away to port with 03 POB 1530 CGNR-6596 commenced touch and go landings 1536 CG6596 off the deck and away to port having completed 04 touch and go landings 1539 CGNR-6596 on deck with Talon engagement 1540 Helo Crash on deck Drill 1541 Helo secured engines and disengaged rotors 1548 SHE-2 (text unreadable) 1550 SHE-2 raised to the main deck rail, coxswain and crew safely aboard. 1555 SHE-2 cradled for sea. 1558 secured Flight Quarters Condition I primary tie downs in place. 1605 Secure from Helo rash on Deck Drill 1750 Set Flight Quarters Condition I for landing, removed secondary tied owns from Helo 1821 Helo off the deck and away to port with 05 POB. 1854 CGNR-6596 on deck. 1835 CGNR-6596 above deck and away to port with 03 POB. 1841 CGNR-6596 on deck above deck and clear to port. 1843 Helo above deck commenced VERTREP with Helo. 1857 Secured VERTREP with CGNR-6596 having transferred 57 gallons of JP-5. 1903 Helo above deck and away to port secured Flight Quarters Condition I set Flight Quarters Condition 2. (part of log unreadable) 2019 secured Flight Quarters Condition I set the Helo Refueling Bill. 2024 Secured from Helo Refueling Bill having transferred 98 gallons of JP-5. 2044 Traversed CGNR-6596 to hanger primary and secondary tie downs in place.

24 Feb – Underway in the Bay of Bengal in position 05° 46.4N 089° 52.6E. Both MGT'S are on the line turning for 27 knots. SHERMAN is under the OPCON and ADCON of COMSEVENTHFLT and TACON of CTU 75.1. 0835

Traversed CGNR-6596 onto the flight deck. Primary and secondary tie downs in place. 0954 Set Flight Quarters Condition 1 for launch. 1023 Helo off deck and away to port with 03 POB. 1025 Secured FLT CON 1, set FLT CON 2. 1132 Set Flight Quarters Condition 1 for landing. 1143 CGNR-6596 is on deck, primary and secondary tie downs in place. 1145 Commence Hot Refuel on deck with Helo. 1150 Secured Hot Refuel having transferred 136 gallons of JP-5 to Helo. 1155 Secured Flight Quarters Condition 1. 1212 Traversed Helo into hanger, primary and secondary tie downs. 1308 Set General Quarters Condition 1 for drill. 1417 Secured from General Quarters Condition 1 drill. 2300 Retarded clocks to conform with -5 Echo time zone. Time now 2200. 2232 Experienced loss of steering casualty, Main Steering Casualty Stations. 2238 Shifted pumps from remote to local. 2239 Energized trick wheel, secured cables. 2246 Test rudder, test sat. 2248 restored steering, no apparent casualty 2249 Engaged standard cable, disengaged trick wheel. 2250 Shifted pumps from local to remote, test rudder, test sat. 2251 secured from steering casualty.

25 Feb – Underway in the Indian Ocean in position 05° 13.2N 082° 48.2E. NR1 MDE is on line turning for 12.5 knots. SHERMAN is under the OPCON and ADCON of COMSEVENTHFLT and TACON of CTU 75.1. 0521 JP-5 test complete sediment 4 mg/l H2O<5PP FSII .18

26 Feb - Underway in the Indian Ocean in position 06° 17.3N 077° 25.6E. NR2 MDE is on line turning for 12.5 knots. SHERMAN is under the OPCON and ADCON of COMSEVENTHFLT and TACON of CTU 75.1. 1430 Queried by Indian warship SUJATA in position 07 30.4N 075 02.7E

1600 Set Flight Quarters Condition 1 for launch. 1608 Traversed CGNR-6596 from hanger, removed primary tie downs 1628 Helo above deck and away to port with 03 POB. 1647 Hailed by Indian Naval aircraft in position 07 38.7N 074 53.4E. Secured Flight Quarters Condition 1, set Flight Quarters Condition 2. 1720 Coxswain and boat crew safely aboard SHE-2. 1724 SHE-2 lowered to the water and away to port with BM1 as coxswain and MK3, SN, SN, SN as boat crew. 1751 Set Flight Quarters Condition one for landing. 1757 CGNR-6596 on deck with primary tie downs in place. 1759 Commenced Hot Refuel on deck with Helo. 1803 Secured Hot Refuel on deck with Helo having transferred 114 gallons of JP-5. 1813 Helo off deck and away to port for training with 03 POB. 1816 Secured from Flight Quarters Condition one. Set Flight Quarters Condition two. 1822 SHE-2 alongside to port. 1825 SHE-2 raised to the port main deck rail coxswain and boat crew safely aboard. 1846 Set Flight Quarters Condition one for landing. 1912 Helo on deck commencing touch and go landings. 1925 Helo on deck with primary tie having completed 06 touch and go landings and 01 landing with Talon engagement. Commenced Hot Refuel on deck with Helo. 1930 Secured Hot Refuel on deck with Helo after having transferred 118 gallons of JP-5. 1950 SHE-2 cradled for sea. 2106 Traversed Helo into hanger primary and secondary tie downs in place.

27 Feb - Underway in the Indian Ocean in position 08° 05.6N 074° 00.6E. NR2 MDE is on line turning for 12.5 knots. SHERMAN is under the OPCON and ADCON of COMSEVENTHFLT and TACON of CTU 75.1. 0840 Commenced Tactical Maneuvering. 0940 Captain Ryan assumed the Conn. 0946 CWO2 Assumed the Conn 1130

Secured from Tactical Maneuvering. 1355 Commenced CIWS PAC Fire. 1405 Secured from CIWS PAC Fire. 1540 Traversed CGNR-6596 on to the flight deck primary tie downs in place. 1630 Set Flight Quarters Condition 1 for launch. 1655 removed primary tie downs from CGNR-6596 Helo off the deck and away to port with 05 POB. 1658 Secured Flight Quarters Condition 1 set Flight Quarters Condition 2. 1730 SET Flight Quarters Condition 1 for landing. 1752 LONE WOLF 43 on deck. 1753 primary tie downs in place 1803 removed primary tie downs from LONE WOLF 43 1804 Helo above deck, Helo on deck. 1805 Helo above deck, Helo on deck. 1806 Helo above deck and away to port. 1854 CGNR-6596 on deck with primary tie downs in place. 1900 Removed primary tie downs Helo away to port 1904 Secured from Flight Quarters Condition 1 set Flight Quarters Condition 2. 1910 SHE-2 lowered to the water with BM3 as coxswain, SN, SN, SN, and MK3 as boat crew. 1914 SHE-2 away to port. 1922 Secured all Navigation Lights to conduct small boat exercise. 2000 Set Flight Quarters Condition 1 for landing, energized all standard Navigation and restricted in ability to maneuver lights 2010 CGNR-6596 on deck with primary tie downs in place. 2013 Commenced Hot Refuel on deck with Helo. 2017 secured Hot refuel with Helo having transferred 149 gallons of JP-5 to Helo. 2024 Secured Flight Quarters Condition 1 SHE-2 alongside to port 2026 SHE-2 raised to the main deck rail 2038 SHE-2 cradled for sea. 2044 Helo hangered wit secondary tie downs in place. 2300 Retarded ships clocks to conform with -4 DELTA time zone, time now 2200 (-4D)

28 Feb – Underway in the Arabian Sea conducting MEF01-1 in position 11° 21.9N 070° 44.0E. NR1 MGT is online turning for

17 knots. NR1 MDE is OOC. SHERMAN is under the OPCON and ADCON of COMSEVENTHFLT and TACON of CTU 75.1. <u>0356</u> Shifted OPCON to COMFIFTHFLT and TACON to CDS 50.1 in position 12 10.0N 070 00.0E <u>1309</u> Commenced CBR and GQI Drill. <u>1530</u> Secured from Drill.

01 Mar – Underway in the Arabian Sea conducting MEF01-1 in position 16° 32.6N 066° 18.3E. NR2 MGT is on line turning for 17 knots. SHERMAN is under the OPCON and ADCON of COMFIFTHFLT and TACON of CTF50. <u>0812</u> Traversed CGNR-6596 on to flight deck, primary and secondary tie downs in place. <u>0845</u> Set Flight Quarters Condition one for launch, removed secondary tie downs from Helo. <u>0921</u> Helo off the deck and away to starboard with 04 POB. <u>0926</u> Secured Flight Quarters Condition 1, set Flight Con 2. <u>1403</u> Set Fight Quarters Condition one for landing. 05 POB. <u>1417</u> Helo on deck with primary tie downs in place. <u>1420</u> Commenced Hot Refuel on deck with Helo. <u>1424</u> Secured Hot Refuel with Helo having transferred (log not readable)

02 Mar - Underway in the Arabian Sea conducting MEF01-1 in position 21° 23.6N 062° 10.3E. NR2 MGT is on line turning for 20 knots. SHERMAN is under the OPCON and ADCON of COMFIFTHFLT and TACON of CTF50. <u>0608</u> USS STETHEM'S small boat STEADFAST, enroute SHERMAN. <u>0620</u> TC2 safely aboard STEADFAST. <u>0611</u> STEADFAST away to port enroute STETHEM. <u>0625</u> STEADFAST enroute SHERMAN. <u>0633</u> STEADFAST alongside to port TC2 safely aboard SHERMAN. <u>1354</u> Traversed CGNR-6596 on to the flight deck, primary and secondary tie downs in place. <u>1418</u> Set Flight Condition 1 for launch, removed secondary tie downs. <u>1441</u> Helo off the deck and away to port with 04 POB enroute

USS PAUL F. FOSTER. 1443 Secured Flight Con 1. 2050 Set Flight Quarters Condition 1 for landing 2119 CGNR-6596 on deck primary tie downs in place. 2122 Commenced Hot Refuel on deck with Helo. 2124 Secured Hot Refuel with Helo having transferred 68 gallons of JP-5. 2125 Secured from Flight Quarters Condition 1. 2255 Traversed Helo into hanger, primary and secondary tie downs in place.

03 Mar – Underway in the Gulf of Oman conducting MEF01-1 in position 025° 33.2N 056° 55.7E. NR1 MGT is online turning for 10 knots. NR1 MDE is OOC. SHERMAN is under the OPCON and ADCON of COMFIFTHFLT and TACON of CTF50. 0615 Small boat from USS STETHEM alongside to port with 05 POB 0616 Small boat away to port with 06 POB including TC2 0620 Small boat from STETHEM alongside to port, TC2 safely aboard. 0626 Small boat away with 05 POB. 1230 Set the UNREP Bill. 1303 Commenced approach on the USS PECOS (TAO-197) 1308 Alongside the USS PECOS for refueling. 1320 Received and coupled fuel hose. 1322 Commenced fueling. 1456 Completed fueling having received 117,155 gallons of JP-5 fuel. 1458 Uncoupled fueling hose and passed to USS PECOS. 1504 Secured the Underway Replenishment Bill. 1513 Set Flight Quarters Condition 1. 1525 Set Flight Con 3 for VERTREP. 1532 PUMA 202 over the deck, commenced VERTREP with Helo. 1600 Secured VERTREP with Helo having transferred 06 pallets of dry and frozen stores. 1601 Helo away to port. 1604 Secured Flight Quarters Condition 3. 2300 Retarded clocks 1 hour to conform to -3 Charlie, time now 2200. 2218 Commenced all preparations for entering port in Mina Salman, Bahrain.

Note – Transited the Strait of Hormuz before refueling with USNS PECOS

04 Mar – Underway in the Persian Gulf conducting MEF01-1 in position 26° 44.0N 051° 40.9E. NR2 MDE is on line turning for 7.5 knots. NR1 MDE is OOC. SHERMAN is under the OPCON and ADCON of COMFIFTHFLT and TACON of CTF50. 0730 Set the Navigation and Anchor Detail. 0832 Set the Special Sea Detail. 0855 Embarked Bahrain Port Control Pilot to port. 0857 Embarked 03 members of U.S. Coast Guard Port Security Unit. 0958 Moored starboard side to Berth 5, Deep Water Jetty, Al Jufayr, Bahrain. 1008 Secured Special Sea Detail. 1025 The OOD shifted the watch from the Pilothouse to the Quarterdeck.

05 Mar – Moored starboard to Berth 5, Deep Water Jetty, Al Jutayr, Bahrain. Ships status is Bravo. The ship receiving potable water, sewage, electricity, and telephone services via shore tie. Liberty has expired for all hands. SHERMAN is SOPA. 0700 Held morning muster for duty sections 1 and 2. All hands present or accounted for. 1300 Liberty granted to all hands with the exception of duty section II to expire 0000 06 Mar 01.

06 Mar - Moored starboard to Berth 5, Deep Water Jetty, Al Jutayr, Bahrain. Ships status is Charlie. 0000 Liberty expired for all hands. 0700 Held morning muster for duty sections 2 and 3. 1316 Liberty granted to all hands with the exception of duty section III to expire 0000 07 Mar 2001.

07 Mar - Moored starboard to Berth 5, Deep Water Jetty, Al Jutayr, Bahrain. Ships status is Alpha. 0000 Liberty expired for all hands. All hands present or accounted for. 0530 Commenced preparations for getting underway. 0534 Set the Special Sea Detail. 0755 Embarked Bahrain Port Control. 0836 Disembarked Bahrain Port Control Pilot. 0817 Underway 0850 Secured Special Sea Detail with the exception of the Nav and Anchor Detail. 0853 Secured Anchor Detail. 0910 Secured Nav Detail

08 Mar - Underway in the Persian Gulf in position 28° 34.0N 50° 02.5E conducting MEF01-1. NR2 MDE is on line turning for 12.5 knots. NR1 MDE is OOC. SHERMAN is under the OPCON and ADCON of COMFIFTHFLT. 0914 Small boat alongside to port. 12 passengers safely aboard SHERMAN. 1040 USS STUMPS small boat alongside to port. 12 passengers safely on board RHIB. 1045 RHIB away to port. 1108 Set Flight Quarters Condition One for launch. * Late entry: 1005 Traversed Helo onto flight deck primary tie downs in place. 1148 Helo off deck and away to starboard 04 POB. 1151 Secured Flight Quarters Condition One set Flight Quarters Condition Two. 1214 Small boat alongside to starboard. 1218 10 passengers safely onboard the small boat. Small boat away to starboard enroute ASIAN BEAUTY. 1234 Small boat alongside to starboard. 1237 Heath and Comfort Team embarked small boat, 5 Sherman personnel, 5 Navy. 1239 Small boat away to starboard enroute to conduct Health and Comfort Boardings. 1314 SHE-2 lowered to the water with BM3 as coxswain, BM2, MK3, SN and SN as crewman. 1315 SHE-2 u/w for training in the mouth of Khowr-E Musa. 1318 Set Flight Quarters Condition 1 for landing. 1332 CGNR-6596 on deck with primary tie downs in place. 1334 Commenced Hot Refuel on deck with Helo. 1338 Secured Hot Refuel on Deck having passed 142 gallons of JP-5 to Helo. 1343 Secured from Flight Quarters. (Log unreadable) small boat enroute USS STUMP. 1355 Small boat alongside USS STUMP. 1358 Small boat away to port. 1431 Helo off deck and away to port w/3POB. 1432 Secured from Flt Con 1, set Flt Con 2. 1440 SHE-2 alongside to port. 1442 SHE-2 raised to the main deck rail and secured for sea, coxswain and boat crew safely aboard. 1503 Receiving small boat alongside to starboard. 1518 Boarding Team embarked SHERMAN. 1519 Small boat away to starboard w/3 POB enroute STUMP. 1534 Set Flt Con 1 for landing. 1546 CGNR-6596 on deck, primary tie downs in place. 1547 Commenced Hot Refuel on deck with Helo. 1551 Secured Hot Refuel on deck having transferred 111 gallons of JP-5 to the Helo. 1555 Secured from Flt Con 1. 1558 Health and Comfort Team safely onboard M/V AMIRA. 1600 SHE-2 Raised and cradled for sea. 1625 Small boat alongside to starboard. 1626 Health and Comfort Team safely embarked SHERMAN. 1628 Small boat away to starboard enroute USS STUMP. 1804 SHE-2 lowered to the main deck rail with SN, SN, MK3 and SN as boat crew and BM1 as coxswain. 1826 SHE-2 away to port wit SN and SN as additional boat crew 1905 SHE-2 alongside to port with boat crew safely aboard SHERMAN. 1907 SHE-2 cradled for sea.

09 Mar – Underway in the Northern Arabian Gulf in position 29° 15.2N 5 49° 07.2E conducting MIO operations. NR2 MDE is online turning for 8.5 knots. NR1 MDE is OOC. SHERMAN is under the OPCON and ADCON of COMFIFTHFLT. 0847 Maneuvering to intercept tanker BERGIE INGERID on course 175 T. 1000 Queried M/V BERGE INGERID. 1057 Set Flight Quarters Condition 1 for launch. 1124 CCGNR-6596 above deck and away to port with 03 POB. 1127 Secured from Flight Quarters Condition 1, set Flight Quarters Condition 2. 1313 Set Flight Quarters Condition One for landing 03 POB. 1324 CGNR-6596 on deck primary tie downs in place. 1326 Commenced Hot Refuel on deck with Helo. 1330 Secured from Hot Refuel on deck with Helo having transferred 157 gallons of JP-5. 1342 Helo off deck away to starboard 04 POB. 1344 Secured from Flight Quarters Condition One, set Flight Quarters Condition Two. 1509 Set Flight Quarters Condition One for landing 04 POB. 1509 Helo on deck primary tie downs in place. (log unreadable) Secured from Hot Refueling on deck with Helo having transferred 130 gallons of JP-5. 1520 Secured from Flight Quarters Condition One.

10 Mar - Underway in the Northern Arabian Gulf in position 29° 43.9N 49° 17.4E conducting MIO operations. NR2 MDE is online turning for 3.5 knots. NR1 MDE is OOC. SHERMAN is under the OPCON and ADCON of COMFIFTHFLT. 1106 Set Flight Quarters Condition 1 for launch. 1130 CGNR-6596 off the deck and away to port with 03 POB. 1131 Secured from Flight Quarters condition III. 1139 DESERT DUCK -744 over the deck. 1140 Commenced VERTREP with Helo. 1146 Secured VERTREP with Helo, having transferred and received cargo and mail. 1150 Helo away to port. 1153 Secured Flight Quarters Condition 3, set Flight Quarters Condition 2. 1317 Set Flight Quarters Condition One. 1325 CGNR-6596 on deck with primary tie downs. 1328 Commenced Hot Refuel on deck with Helo. 1332 Secured Hot Refuel on deck having transferred 126 gallons of JP-5 to CGNR-6596. 1358 Helo off deck and away to port with 4 POB. 1349 Secured Flight Quarters Condition One, set Flight Quarters Condition Two. 1515 Set Flight Quarters

Condition One for landing. 1526 CGNR-6596 on deck, primary tie downs in place. 1529 Commenced Hot Refuel on deck with Helo. 1533 Secured Hot Refuel on deck having transferred 147 gallons of JP-5 to Helo. 1539 Secured Flight Quarters. 1705 Traversed Helo to hanger, primary and secondary tie downs in place.

11 Mar - Underway in the Northern Arabian Gulf in position 29° 36.4N 49° 11.0E conducting MIO operations. NR2 MDE is online turning for 8.5 knots. NR1 MDE is OOC. SHERMAN is under the OPCON and ADCON of COMFIFTHFLT. 1039 Traversed CGNR-6596 onto flight deck, primary and secondary tie downs in place. 1109 Set Flight Quarters Condition One for launch, removed secondary tie downs from Helo. 1132 Helo off deck and away with 04 POB. 1149 CGNR-6596 reported excessive vibration, set Flight Quarters Condition One for landing, 04 POB. 1154 Helo on deck with primary tie downs in place. 1208 Secured from Flight Quarters Condition One. 1341 Set Flight Quarters Condition One for launch, removed secondary tie downs from Helo. 1400 Helo off deck and away to starboard, 03 POB. 1404 Helo on deck with primary tie downs in place. 1406 Commenced Hot Refuel on deck with Helo. 1409 Secured Hot Refuel on deck having transferred 70 gallons of JP-5 to Helo. 1410 Embarked 01 person on to the Helo. 1412 Helo off deck and away to starboard, 04 POB. 1550 Set Flight Quarters Condition One for landing, 04 POB. 1603 Helo on deck primary tie downs in place. 1605 Commenced Hot Refuel on deck with Helo. 1610 Secured Hot Refuel on deck having transferred 145 gallons JP-5 to Helo. 1615 Secured from Flight Quarters Condition One. 1625 SHE-2 raised and cradled for sea. 1810 Helo hangered. 2300 Secured all standard navigation lights by order of the Commanding Officer.

12 Mar - Underway in the Northern Arabian Gulf in position 29° 41.8N 48° 59.2E conducting MIO operations. NR2 MDE is online turning for 8.5 knots. NR1 MDE is OOC. SHERMAN is under the OPCON and ADCON of COMFIFTHFLT. 0633 Set Flight Quarters Condition One for launch, removed secondary tie downs from CGNR-6596. 0705 Helo off the deck and away to port w/4POB. 0709 Secured from Flight Quarters Condition One, set Flight Quarters Condition 2. 0838 Set Flight Quarters Condition One for landing. 0850 CGNR-6596 on deck with primary tie downs engaged. 0852 Commenced Hot Refuel on deck with Helo. 0855 Secured Hot Refuel having transferred 133 gallons of JP-5 to Helo. 0900 Secured from Flight Quarters Condition One. 1105 Set Flight Quarters Condition One for launch removed secondary tie downs. 1119 Helo off deck and away to starboard with 3 POB. 1121 Secured Flight Quarters Condition One, set Flight Quarters Condition 2. 1257 Set Flight Quarters Condition One for landing. 1310 CGNR-6596 on deck with primary tie downs in place. 1312 Commenced Hot Refuel on deck wit Helo. 1316 Secured Hot Refuel on deck having transferred 140 gallons of JP-5 to Helo. 1321 Secured from Flight Quarters Condition One. 1435 Set the Underway Replenishment Bill. 1504 Commenced approach on USNS KANAWA (TAO-196) 1507 Alongside USNS KANAWA 1510 Traversed Helo into hanger, primary and secondary tie downs in place. 1523 Received and coupled fuel hose. 1528 Commenced fueling 1605 Completed fueling having received 41, 988 gallons of JP-5 fuel 1605 Detensioned spanwire, uncoupled fuel hose and passed to USNS KANAWA 1607

Passed spanwire from USS KANAWA. <u>1612</u> Secured the Underway Replenishment Bill. <u>1927</u> Commenced preparations for entering port, Bahrain.

13 Mar - Underway in the Northern Arabian Gulf in position 27° 08.6N 51° 14.0E conducting MIO operations. NR2 MDE is online turning for 9 knots. SHERMAN is under the OPCON and ADCON of COMFIFTHFLT. <u>0630</u> Set the Navigation and Anchor Detail. <u>0707</u> Pilot boat alongside to port. <u>0806</u> Moored STBD side to berth 5 Deep Water Jetty, Mina Salman Bahrain. <u>0809</u> Secured Special Sea Detail. 0840 Duty Section One is on watch. <u>1130</u> Liberty granted to duty section 2 and 3 to expire onboard for duty section 2 0845 14 Mar 01 and 0845 15 Mar 01 for duty section 3.

14 Mar – Moored Starboard side to Berth 5 Deep Water Jetty, Mina Salman Bahrain with standard mooring lines doubled. Under the OPCON and ADCON of COMFIFTHFLT. Ships status is Charlie. All deck, anchor and aircraft warning lights are energized and burning brightly. Material Condition Yoke is set throughout the ship. Receiving potable water and sewage via barge ties. Receiving telephone services via shore ties. All hands are on authorized liberty with the exception of duty section 1. Other Naval ships are present. <u>0925</u> Liberty granted to duty section 1. <u>2300</u> (rate and name redacted) reported onboard from CG MLC Pacific.

15 Mar - Moored Starboard side to Berth 5 Deep Water Jetty, Mina Salman Bahrain with standard mooring lines doubled. Under the OPCON and ADCON of COMFIFTHFLT. Ships status is Charlie. All deck, anchor and aircraft warning lights are energized and burning brightly. Material Condition Yoke is set throughout the ship. Receiving potable water and sewage via barge ties. Receiving telephone services via shore ties. All hands are on authorized liberty with the exception of duty section 2. <u>0000</u> Liberty expired for all E-4 and below. All hands are present or accounted for. <u>0845</u> Liberty expired for duty section 3. Held morning muster for duty section 2 and 3. All hands present or accounted for. <u>0944</u> Liberty granted to duty section 2 to expire no later than 0000 (log unreadable) <u>1315</u> B.A.N.S SABHA underway - Bahrainy Navy FFG-90. <u>1945</u> CWO2 arrived PCS from MLC PAC <u>2213</u> Commenced preparations for getting underway. 2400 Liberty expired for all hands.

Note – RBNS SABHA (FFG-90) used to be the USS JACK WILLIAMS (FFG-24) and given to Bahrain.

16 Mar - Moored Starboard side to Berth 5 Deep Water Jetty, Mina Salman Bahrain with standard mooring lines doubled. Under the OPCON and ADCON of COMFIFTHFLT. Ships status is Alpha. All deck, anchor and aircraft warning lights are energized and burning brightly. Material Condition Yoke is set throughout the ship. Receiving potable water and sewage via barge ties. Receiving telephone services via shore ties. All hands are on authorized liberty with the exception of duty section 3. <u>0830</u> Liberty expired for all hands. <u>0835</u> Water barge away to port. <u>0930</u> Set the Special Sea Detail. <u>0956</u> Embarked Bahrain Port Control Pilot. <u>1004</u> Underway from Mina Salman, enroute the North Arabian Gulf. <u>1027</u> Bahrain Pilot vessel alongside to starboard. <u>1038</u> Bahrain Port Control Pilot Disembarked Sherman, safely aboard pilot vessel, pilot vessel away to starboard. <u>1040</u> Secured the Special Sea Detail with the exception of Navigation and Anchor Detail. 1046 Secured the Anchor Detail. 1058 Secured the Navigation Detail.

17 Mar – Underway in the Persian Gulf conducting MIO Ops. Both MDE'S are online turning for 16.5 knots. SHERMAN is under the OPCON and ADCON of COMFIFTHFLT. 0437 Commenced query of vessel in position 29° 48N 49° 18.6E 044 Vessel identified to be M/V AMROZ, under Syrian flag. Commenced escorting vessel to anchorage point 29° 14N 49° 16E. 0640 Set Flight Quarters Condition One for launch. 0638 Late entry – Traversed Helo onto flight deck primary and secondary tie downs in place. 0704 Helo off deck and away to port, 04 POB. 0707 Secured from Flight Quarters Condition One, set Flight Quarters Condition Two. 1018 Commenced query of M/V AL JUZUF in position 29° 41.3N 049° 04.0E on course 100° True, under flag. 1050 Commenced query of M/V in position 29° 41.5N 03.6E on course 285° True, vessel identified to be M/V SENOBAR, under Iranian flag. 1148 Set Flight Quarters Condition One. 1158 CGNR-6596 is on deck with primary tie downs in place. 1200 Commenced Hot Refuel on deck with Helo. 1204 Secured Hot Refuel on deck having transferred 90 gallons of JP-5 to Helo. 1209 Helo above deck and away to port with 03 POB. 1257 Set Flight Quarters Condition One for landing. 1309 Helo is on deck with primary tie downs in place. 1313 Secured Flight Quarters Condition One, set the Helo Refueling Bill. 1318 Secured the Refueling Bill after having transferred 88 gallons of JP-5 to Helo. 1338 Installed secondary tie downs on Helo. 1843 Commenced query of vessel in position 29° 26N 048° 55E on course 164° True, vessel identified to be M/V LUBAN under Chinese flag. 1903 Secured all standard navigation lights upon order of the

Commanding Officer. 2005 Traversed CGNR-6596 into hanger, primary, secondary and heavy weather tie downs in place. 2038 Observed Dhow in position 29° 33.5N 01.5E throwing nets into the water. 2320 Commenced query of vessel in position 28° 40.5N 048 57.1 E on course 114° True. 2326 Energized all standard navigation lights. 2343 Vessel identified as M/V SIGIRI under Sri Lankan flag.

18 Mar - Underway in the North Arabian Gulf in position 29° 39.1N 48° 55.3E conducting MIO Ops. Both MDE'S are online turning for 10 knots. SHERMAN is under the OPCON and ADCON of COMFIFTHFLT. 0259 Commenced query of vessel in position 29° 44.3N 048° 57.8E on course 068° True. 0557 Maneuvered to intercept contact in position 29° 50.0N 049° 13.6E on course of 050° True. 0630 Entered Iranian claimed Territorial Sea in position 29° 44.99N 049° 13.0E. 0620 On scene with M/V AL MAHA in position 29° 50.8N 049° 14.8E. 0649 Broke off pursuit of M/V AL MAHA. 0718 Set Flight Quarters Condition One for launch. Removed secondary tie downs from Helo. 0720 Set the Helo Refueling Bill. 0725 Commenced Cold Refuel on deck with Helo. 0726 Secured Cold Refuel on deck with Helo having transferred 55 gallons of JP-5. 0754 Helo off deck and away to starboard with 04 POB. 0759 Secured from Flight Quarters Condition One, set Flight Quarters Condition 2. 0943 Set Flight Con 1 for landing. 0959 CGNR-6596 on deck, primary tie downs in place. 1007 Commenced Hot Refuel on deck with Helo. 1004 Secured Hot Refuel having transferred 81 gallons of JP-5 to CGNR-6596. 1016 Removed primary tie downs. 1017 CGNR-6596 above deck and away to port with 06 POB. 1022 Secured from Flight Quarters Con 1, set Flight Con 2. 1133 Commenced VERTREP with US Navy Helo Desert Duck. 1138 Secured VERTREP having transferred 27 pounds of mail. 1233 Set Flight Con 1 for landing. 1230 Entered the ICTW in position 29° 40.5N 49° 34.6 E on course 350° True at 5 knots. 1244 Helo is on deck, primary tie downs in place 1246 Commenced Hot Refuel on deck with Helo. 1251 Secured the Hot Refuel having transferred 149 gallons of JP-5 to Helo. 1255 Secured Flight Con 1. 1436 Set Flight Con 1 for launch. 1505 Helo off deck an away to port with 04 POB. 1507 Secure Flight Con 1, set Flight Con 2. 1522 Set Flight Con 1 for landing. 1536 Helo is on the deck, primary tie downs in place. 1539 Secured Flight Con 1. 1545 Installed secondary tie downs on Helo. 1825 CGNR-6596 Traversed to the hanger. 2125 Crossed back over Iranian Claimed Territorial Water line in position 29° 20.6N 049° 52.0E.

Note – ICTW is short for Iranian Claimed Territorial Water.

19 Mar - Underway in the North Arabian Gulf in position 27° 44.4N 49° 18.6E conducting MIO Ops. NR2 MDE is online turning for 3.5 knots. SHERMAN is under the OPCON and ADCON of COMFIFTHFLT. 0612 Traversed Hel onto flight deck with primary tie downs in place. 0635 Set Flight Quarters Condition One for launch. (Logs unreadable) 0658 Secured Flight Quarters Condition One 1046 Set Flight Quarters Condition One for landing. 1056 Helo is on deck with primary tie downs in place. 1057 Commenced Hot Refuel on deck with Helo. 1103 Secured Hot Refuel on deck having transferred 171 gallons of JP-5 to the Helo. 1106 Secured Flight Quarters Condition One. 1205 Crossed Line Of Demarcation for Iranian Claimed Territorial Sea Limit. 1330 Crossed into International Territorial Water in position 29° 39.3N 049° 35.5E. 1815 Secured all standard navigation lights by order of the

Commanding Officer. 1917 Energized all standard navigation lights. 1927 alongside vessel in position 29° 40.1N 049° 01.0E. Commenced query. 1941 vessel identified as M/V DIAMOND under Honduras flag. 1947 Broke off query of M/V DIAMOND with no response. 1957 Commenced query of vessel in position 29° 40.1N 049° 00.8E, vessel identified to be M/V AGON. 2018 Directed M/V AGON to anchorage NR4, 29° 14N 049° 16E. 2105 Secured all standard navigation lights by order of the Commanding Officer. 2236 Commenced query of vessel in position 29° 43.1N 049 12.6E on course 087°T vessel identified to be M/V HANEN. 2242 Directed HANAN to anchorage NR5 29° 14N 049 20.E

Note – USS STETHEM would later bust M/V DIAMOND carrying 7,462 tons of illegal oil, which was the 3rd largest bust at the time. In their command history report after the deployment they stated that "The seizure of this oil prevented Saddam Hussein from making 2 million dollars."

20 Mar - Underway in the North Arabian Gulf in position 29° 35.4N 50° 15.0E conducting MIO Ops. Both MDE'S are online turning for 5 knots. SHERMAN is under the OPCON and ADCON of COMFIFTHFLT. 0550 Commenced query of vessel in position 29° 44.3N 49° 15.4E on course 281° True. Vessel unidentified. 0612 Traversed Helo onto flight deck, primary and secondary tie downs in place. 0635 Set Flight Con 1 for launch. (Logs unreadable) 0705 secured from Flight Con 1. 1026 Set Flight Con 1 for landing. 1036 Helo is on deck, primary tie downs in place. 1032 Commenced Hot Refuel on deck with Helo. 1042 Secured Hot Refuel having transferred 144 gallons of JP-5 to Helo. 1047 Secured Flight Con 1. 1102 Installed secondary tie downs to Helo. 1403 Set Flight Quarters

Condition One for launch removed secondary tie downs from Helo. 1418 Helo off deck and away to port, 04 POB enroute USS HARRY S. TRUMAN to transfer RD3 Thomas Ryan for medical evaluation. 1421 Secured from Flight Con One, set Flight Quarters Condition Two. (log unreadable) Set Flight Quarters Condition 1 for landing 05 POB. 1643 Helo on deck with TALON engagement, 02 passengers disembarked. 1644 Commenced Hot Refuel on deck with Helo. 1647 Secured Hot Refuel on deck having transferred 189 gallons of JP-5 to Helo. 1652 Secured from Flight Con One, primary and secondary tie downs installed on Helo. 1839 SHE-2 lowered to the main deck rail. 1841 Boat crew embarked SHE-2, boat crew consists of BM2, BM3, MK3, SN. 1843 SHE-2 lowered to the water. 1844 SHE-2 away to port w/3 POB enroute USS PAUL F FOSTER. 1905 SHE-2 alongside to port, passenger's and mail safely embarked SHERMAN. 1908 SHE-2 raised to main deck rail. 1911 SHE-2 lowered to the water, away to port. 2036 SHE-2 away to port enroute USS PAUL F FOSTER. 2105 SHE-2 alongside to port. 2120 SHE-2 Cradled and secured for sea.

21 Mar - Underway in the North Arabian Gulf in position 29° 38.4N 49° 05E conducting MIO Ops. NR2 MDE is online turning for 8.5 knots. SHERMAN is under the OPCON and ADCON of COMFIFTHFLT. 1507 Commenced query of vessel in position 29° 26.3N 049° 10.6 E on course 145° true. 1545 Vessel identified to be M/V TOOE KAREM.

22 Mar - Underway in the North Arabian Gulf in position 29° 37.8N 49° 06.3E conducting MIO Ops. SHERMAN is dead in the water. Both MDE'S are in immediate standby. All standard lights are secured by order of the Commanding Officer. SHERMAN is under the OPCON and ADCON of

COMFIFTHFLT. 0422 Traversed CGNR-6596 from the hanger 0425 Set Flight Quarters Condition 1 for launch. 0457 Helo above deck and away to port with 03 POB. 0500 Secured from Flight Quarters Condition 1, set Flight Quarters Condition 2. 1021 Set Flight Quarters Condition One for Landing, 04POB. 1028 Helo on deck with TALON engagement. 01 passenger disembarked the Helo. 1029 Commenced Hot Refuel on deck with Helo. 1030 SHE-2 lowered to the main deck rail. 1034 Secured Hot Refuel on deck after having transferred 150 gallons of JP-5 to helo. 1039 Secured from Flight Con 1. 1042 SHE-2 cradled for sea. 1115 Maneuvered to intercept contact vessel in position 29° 16.9N 049° 24.2E. 1150 Commenced query of vessel in position 29° 20N 049° 20E on course 306°T at 5 knots. Vessel identified to be M/V HASSAN 1 under SAO TAME flag. 1247 Set Flight CON 1 for launch. 1307 Helo is off the deck and away to port, with 03 POB. 1308 Secured Flight Con 1, set Flight Con 2. 1353 SHE-2 lowered to the main deck rail. 1417 SHE-2 lowered to the water. (log unreadable) 1433 SHE-2 alongside USS FOSTER. 1434 Captain Ryan and CDR safely aboard USS FOSFTER. 1500 Set Flight Quarters Condition 1 for landing. 1512 Helo is on deck, primary tie downs in place. 1513 Commenced Hot Refuel on deck with Helo. 1516 Secured Hot Refuel having transferred 175 gallons of JP-5 to Helo. 1517 SHE-2 enroute SHERMAN. 1522 Secured Flight Con 1. 1523 SHE-2 alongside to port. 1524 SHE-2 raised to the main deck rail. 1525 coxswain and boat crew safely aboard SHERMAN. 1612 USS FOSTERS small boat enroute SHERMAN with Capt. Ryan. 1618 Capt. Ryan safely aboard SHERMAN. 1716 Traversed CGNR-6596 into hanger, primary, secondary and heavy weather tie downs in place. <u>1755</u> Observed sunset, all standard navigation lights secured by order of the Commanding Officer.

23 Mar - Underway in the North Arabian Gulf in position 29° 33.1N 49° 16.2E conducting MIO Ops. NR1 MDE is online turning for 8.5 knots. All standard lights are secured by order of the Commanding Officer. SHERMAN is under the OPCON and ADCON of COMFIFTHFLT. <u>0232</u> US Navy SEAL boat alongside to port embarked 01 POB. <u>0233</u> US Navy SEAL safely aboard SHERMAN, small boat away to port <u>0252</u> SEAL team RHIB alongside to starboard <u>0256</u> Embarked Navy SEAL team. <u>0300</u> 02 Navy SEAL RHIB's moored alongside to starboard. <u>0412</u> US Navy SEAL RHIB'S away to starboard. <u>0414</u> Maneuvering to intercept M/V in position 29° 46.2N 049° 05.9E. <u>0520</u> 02 Navy SEAL RHIB's alongside to starboard. <u>0523</u> Navy SEAL RHIB's away to starboard. <u>0443</u> Late Entry – M/V identified to be M/V YEMAYA under BELIZE flag. <u>0610</u> SHE-2 lowered to main deck rail. <u>0635</u> Set Flight Con 1 for launch. <u>0615</u> Late Entry –Removed primary and secondary tie downs from CGNR – 6596, traversed from hanger <u>0653</u> SHE-2 away to port with BM3 as coxswain BM3 and MK3 as boat crew and Boarding Team Blue consisting of CWO, MK3, SK3, RD2, GM3, SN, FN and LTJG as boarding Officer. <u>0700</u> Helo above deck and away to port. <u>0703</u> Secured Flight Con 1, set Flight Con 2. <u>0710</u> SHE-2 alongside YEMAYA with Boarding Team Blue safely aboard. <u>0840</u> Set Flight Quarters Condition One for landing 03POB. <u>0848</u> Helo on deck with TALON engagement. <u>0850</u> Commenced Hot Refuel on deck with Helo. <u>0854</u> Secured Hot Refuel on deck having transferred 130 gallons of JP-5 to Helo. 0901 Helo off deck and away to port 04 POB. <u>1044</u> Set Flight Quarters Condition 1 for landing 04 POB.

1102 Helo on deck with primary tie downs. **1104** Commenced Hot Refuel on deck with Helo. **1108** Secured Hot Refuel on deck after having transferred 150 gallons of JP-5 to Helo. **1114** Secured from Flight Con One. **1155** Boarding of M/V YEMAYA completed. **1156** Boarding Team Blue safely on board SHE-2, enroute SHERMAN. **1158** SHE-2 alongside to port. **1159** 08 members of Boarding Team Blue safely on board SHERMAN. **1201** Coxswain and boat crew safely aboard SHERMAN. **1201** SHE-2 raised to the main deck rail. **1212** SHE-2 cradled and ready for sea. **1325** Set Flight CON 1 for launch. **1341** Helo is off the deck and away to port with 03 POB **1342** Secured Flight Con 1, set Flight Con 2. **1520** Set Flight Con 1 for landing. **1523** Helo is on the deck, primary tie downs in place. **1530** Commenced Hot Refuel on deck with Helo. **1535** Secured Hot Refuel having transferred 148 gallons of JP-5 to Helo. **1536** on scene with M/V GEM and M/V AL DEBARAN in position 29° 20N 049° 13E. **1539** Secured Flight Con 1. **1721** Traversed CGNR-6596. **1917** M/V AL DERBARAN crossed Iranian Territorial Seas in position 29° 39.7N 048° 46.25E on course 300 True at 9.5 knots. **2148** M/V GEM crossed into Iranian Territorial Seas in position 29° 41.2N 048° 49.2E on course 315T at 5 knots. **2200** Lost radar contact M/V AL DEBARAN. **2215** Secured all standard Navigation Lights by order of Commanding Officer.

24 Mar – Underway in the Arabian Gulf in position 29° 40.1N 049° 00.7E conducting MIO OPS. SHERMAN is DIW. Both MDE'S are in standby, both MGT's are in standby, NR1 SSDG is providing electrical power. All standard navigations lights are secured by order of the Commanding Officer. Dog Zebra and material condition Yoke is set throughout the ship, with

EMCON DELTA in effect. SHERMAN is under the OPCON and ADCON of COMFIFTHFLT. 0510 Crossed over the Iranian Claimed Territorial Sea Limit. 0612 Set Flight Quarters Condition 1 for launch. 0630 Helo above deck and away to starboard with 04 POB. 0639 SHE-2 lowered to main deck rail 0700 SHE-2 away to port with BM2 as coxswain, MK3 and BM3 as boat crew and Boarding Team Black consisting of (part of log unreadable) RD3, GM1, SN, FN, MK1, and LTJG as boarding officer 0750 set Flight Quarters Condition One for landing 04 POB. 0804 Helo on deck with primary tie downs. 0806 Commenced Hot Refuel on deck with Helo. 0810 Secured Hot Refuel on deck after having transferred 130 gallons of JP-5 to Helo. 0817 Secured from Flight Quarters Condition One. SHE-2 alongside to port. 0822 Boarding Team Black safely on board SHERMAN. 0823 SHE-2 raised to the main deck rail coxswain and boat crew safely onboard. 0852 SHE-2 cradled and ready for sea. 0926 Crossed over the Iranian Claimed Territorial Water Line in position 29° 40.2N 049° 34.75E. 1135 Traversed CGNR-6596 in the hanger, primary and secondary tie downs in place. 1317 Set the Underway Replenishment Bill. (Part of log unreadable) 1410 Passed bridge to bridge phone line to USNS PECOS. 1412 USNS PECOS passed spanwire. 1417 tensioned spanwire. 1419 Received and coupled Fuel Hose. 1425 Commenced Fueling 1442 Passed Highline. 1445 Tensioned Highline. 1456 Commenced Cargo Transfer 1507 Completed Fueling having received 53,993 gallons of JP-5 fuel. 1512 Uncoupled fuel hose and passed to USNS PECOS. 1604 Completed cargo transfer having received 16 pallets of stores. 1606 De-tensioned Highline. 1610 De-tensioned spanwire. 1611 Returned highline to USNS PECOS. Returned Messenger to USNS PECOS. 1613 Recovered Bridge to bridge phone. Broke away from USNS PECOS (T-AO – 197). 1616 Secured the Underway Replenishment Bill. 1917 Lowered SHE-2 lowered to the main deck rail. 1945 SHE-2 lowered to the water with BM3 as coxswain with SN and MK3 as boat crew. 1946 SHE-2 Away to port enroute USS PAUL F. FOSTER. 2009 SHE-2 alongside to port. 2011 SHE-2 away to port enroute USS PAUL F FOSTER. 2023 SHE-2 alongside to port. 2025 SHE-2 raised to the main deck rail with coxswain and boat crew safely aboard 2040 SHE-2 cradled and ready for sea.

25 Mar – Underway in the North Arabian Gulf in position 29° 42N 49° 10.2 E conducting MIO OPS. Ships course is 257° True turning for 8.5 knots with NR1 MDE clutched in NR2 MDE in standby. Sherman is under the OPCON and ADCON of COMFIFTHFLT. 0453 Clutched in NR2 MDE. 0515 Clutched in both MGT's placed both MDE's in standby maneuvering to intercept M/V in position 29° 48.2N 49° 20.5 E 0520 Crossed over Iranian Claimed Territorial Sea Limit. 0535 Commenced query of vessel in position 29° 49.4N 49° 17.4 E on course at 040° True 0538 SHE-2 lowered to the main deck rail (part of log unreadable) VBSS Black Team consists of LTJG as Boarding Officer and GM1, SN, SN, RD2, MKC, MK1, and MK3. 0601 M/V ALA BAS crossed Iranian Recognized Territorial Water. 0613 Paralleled both SSDG's. 0617 Commenced query on vessel in position 29° 51.9N 49° 13 E on course 330° True. 0653 SHE-2 alongside to port. 0655 SHE-2 raised to the main deck rail with coxswain and boat crew safely aboard. 0705 decluthed both MGT's clutched in both MDE's. 0710 Set Flight Quarters Condition One for launch. 0647 Late Entry – Traversed Helo onto flight deck, primary and secondary tie downs in place. 0731 Helo off deck and away to starboard 04 POB. 0734 Secured from Flight Quarters Condition One, set Flight Quarters Condition Two. 0751 Crossed Iranian Claimed Territorial SEA Limit and into International waters in position 29° 41.5N 49° 15.28 E 0919 Crossed Iranian Claimed Territorial Water Limit in position 29° 38.9N 49° 35.6 E 1206 Set Flight Con 1 for landing. 1213 CGNR-6596 is on the deck, primary tie downs in place, disembarked 01 passenger. 1215 Commenced Hot Refuel on deck with Helo. 1219 Secured Hot Refuel, having transferred 153 gallons of JP-5 TO Helo. 1224 Secured Flight Con 1. 1354 Traversed Helo into the Hanger, primary and secondary tie downs in place.

26 Mar - Underway in the North Arabian Gulf in position 29° 40.2N 49° 18.4 E conducting MIO OPS. Ships course is 290° True turning for 15 knots with both MDE's clutched in. Sherman is under the OPCON and ADCON of COMFIFTHFLT. 0325 Commenced maneuvering to intercept M/V n position 29° 46.1N 49° 25.7 E 0405 Commenced query of M/V SWISSCO 0445 Crossed over Iranian Claimed Territorial Sea Limit. 0630 Traversed CGNR-6596 from hanger. 0710 Set Flight Quarters Condition One for Launch. 0728 Helo away to port with 04 POB. 0733 Secured Flight Quarters Condition One.0756 PAUL F FOSTER small boat alongside to port. 0757 03 POB safely aboard SHERMAN including PAUL F FOSTER's Commanding Officer. 0758 PAUL F FOSTER's small boat away to port. 0812 SHE-2 lowered to the main deck boat rail. 0833 SHE-2 lowered to the water with BM1 as coxswain and SN, MK3 and VBSS Gold Team as boat crew. Gold Team consists of FT1, BM2, RD3, GM3, SN, SN, LTJG and LTJG as Boarding Officer. 0837 SHE-2 away to port. 0848

PAUL F FOSTER small boat alongside to port. 0922 PAUL F FOSTER small boat alongside to port. Navy RHI away to port with 07 POB. 0927 Set Flight Quarters Condition One for landing. 0940 Helo on deck with primary tie downs in place. 0943 Commenced Hot Refuel on deck. (part of log unreadable), 1019 SHE-1 lowered to the main deck rail. 1042 SHE-1 is lowered to the water with BM3 as coxswain, SN, MK3, and VBSS Blue Team as boat crew. Blue Team consists of GM3, RD2, SK2, SN, FT2, SN and ENS. 1043 SHE-1 away to starboard. 1044 SHE-2 alongside to starboard. 1045 SHE-2 away to starboard with BM1 As coxswain, SN, MK3 and remaining Blue Team as boat crew. Blue Team consists of BM2, MK2, FN, SN, and CWO. 1126 SHE-2 alongside to port. 1129 SHE-2 away to port with Gold Team safely aboard. 1130 SHE-1 alongside to port. 1131 SHE-2 away to port with remaining Gold Team safely aboard. 1145 SHE-2 alongside to port. 1147 SHE-2 raised to the main deck rail. 1157 Set Flight Quarters Condition One for landing. 1206 CGNR-6596 on deck with primary tie downs in place. 1208 Commenced Hot Refuel on deck with Helo. 1213 Secured Hot Refuel having passed 155 gallons of JP-5. 1217 Secured Flight Quarters Condition One. 1342 Small boat alongside to starboard. 1342 Two passengers safely embarked small boat. 1344 small boat away to starboard enroute PAUL F FOSTER. 1400 SHE-2 lowered to the rail. 1401 SHE-2 lowered to the water with BM1 as coxswain SN and MK3 as boat crew. 1403 SHE-2 away to port enroute M/V YICK LEE to pick up boarding team Blue. 1405 Completed boarding on M/V YICK LEE. 1409 Boarding Team departed M/V YICK LEE, safely on board SHE-1. 1410 SHE-1 departed M/V YICK LEE enroute SHERMAN. 1412 SHE-1 alongside to starboard, raised to the 01 deck rail, boarding team and boat crew safely aboard SHERMAN. 1415 SHE-1 cradled and secure for sea. 1423 SHE-2 alongside to port with the rest of the boarding team. 1425 Boarding team safely embarked SHERMAN. 1426 Boat crew safely embarked SHERMAN. 1427 SHE-2 raised to the 01 deck rail and cradled for sea. 1428 No violations or warnings were issued to the M/V YICK LEE. 1445 Traversed CNR-6596 into the hanger primary and secondary and heavy weather tie downs in place.

27 Mar – Underway in the North Arabian Gulf in position 29° 41.5N 048° 58.8E conducting MIO OPS. NR1 MDE is on line turning for 8.5 knots. Dog Zebra and material condition Yoke are set throughout the ship with EMCON DELTA in effect. SHERMAN is under the OPCON and ADCON of COMFIFTHFLT. 0625 Traversed Helo onto flight deck with primary and secondary tie downs in place. 0640 Set Flight Quarters Condition One for launch. 0700 Helo is off the deck and away to port. 0705 SHE-1 lowered to the main deck rail. 0725 SHE-1 lowered to the water with BM1 as coxswain SN, MK3, and Gold Team 1 as boat crew. Gold Team consists of SN, SN, CWO, BM2, FT1, RD3, GM3, and LTJG. 0730 SHE-1 away to starboard. 0735 SHE-2 lowered to the water with BM3 as coxswain, SN, MK3, and Gold Team 2 as boat crew. Gold Team 2 consists of FN, SN, MK1, SN, SN, and LTJG. 0739 SHE-2 away to port. 0757 SHE-2 alongside to port. 0800 SHE-2 away to port with BM3 as coxswain, SN, MK3 and FSC as boat crew enroute PAUL F FOSTER. 0833 SHE-2 alongside to port. 0835 SHE-2 raised to the main deck rail with coxswain and boat crew safely aboard. 0900 Set Flight Quarters Condition One for landing. 0910 Helo on deck. 0912 Commenced Hot Refuel on deck with Helo. 0919 Secured Hot

Refuel on deck after having transferred 149 gallons of JP-5 to Helo. 0921 Secured Flight Quarters Condition One. (Part of log unreadable) And RD3 as boat crew. 1027 SHE-1 alongside to starboard. 1028 SHE-1 away to starboard. 1110 SHE-1 alongside to starboard. 1111 SHE-1 away to starboard with BM1 as coxswain, SN, MK3 and MK3 as boat crew enroute PAUL F FOSTER. 1133 Set Flight Quarters Condition One for launch. 1150 Helo off deck and away to port with -4 POB. 1153 Secured Flight Quarters Condition One set Flight Quarters Condition Three for HIFR. 1208 Desert Duck over the deck. 1209 Commenced VERTREP with Helo. 1210 Helo away to port. 1214 Desert Duck over deck. 1221 Helo away to port. 1225 Secured from VERTREP, set Flt Con 2. 1225 SHE-1 alongside to starboard. 1227 01 passengers safely onboard. 1230 SHE-1 away to starboard enroute PAUL F FOSTER. 1310 SHE-1 received switch out of coxswain BM3, SN, MK3. 1322 SHE-1 enroute Paul F FOSTER. 1344 Set Flight Quarters Condition 1. 1352 Helo on deck with primary tie downs installed. 1353 Commenced Hot Refuel on deck with Helo. 1358 Secured Hot Refuel having transferred 156 gallons on JP-5. (No time given) SHE-1 alongside to starboard. 1600 SHE-1 raised to the 01 deck rail, boat crew safely on board SHERMAN. 1603 SHE-1 cradled and secured for sea. 1608 SHE-2 lowered to the main deck rail 1615 boat crew embarked SHE-2. 1620 SHE-2 lowered to the water, coxswain is BM3 crewmen are SN, DC3, MK3. 1623 SHE-2 away to port enroute M/V TAMA TIKI to deliver dinner. 1723 SHE-2 alongside to port. 1725 SHE-2 away to port to retrieve boarding team. 1730 SHE-2 alongside M/V TAMA TIKI. 1736 SHE-2 away from M/V TAMA TIKI enroute with 07 POB of Boarding Team Gold. 1740 SHE-2 alongside to port. 1741 07 boarding team members safely on board SHERMAN. 1742 SHE-2 away to port enroute M/V TAMA TIKI. 1747 SHE-2 alongside M/V TAMA TIKI 1755 SHE-2 away enroute SHERMAN with 07 boarding team members safely onboard. 1756 SHE-2 alongside to port. 1800 SHE-2 away to port. 07 boarding team members safely on board SHERMAN. 1803 SHE-2 alongside M/V TAMA TIKI. 1806 SHE-2 away enroute SHERMAN. (Part of log unreadable) Safely on board SHERMAN. 1824 SHE-2 cradled for sea.

28 Mar – Underway in the North Arabian Gulf in position 29° 43.4N 045 12.4E conducting MIO OPS. NR2 MDE is online turning for 8.5 knots. Dog Zebra and material condition Yoke are set throughout the ship with EMCON DELTA in effect. SHERMAN is under the OPCON and ADCON of COMFIFTHFLT. 0652 SHE-2 lowered to the main deck rail. 0704 SHE-1 lowered to the main deck rail. 0707 SHE-2 lowered to the water with BM1 as coxswain, SN, MK3 and Gold Team 1 as boat crew. Gold Team consists of GM3, RD3, GM1, FT1, SN, CWO, and LTJG. 0711 SHE-2 away to port. (Time not readable) SHE-1 lowered to the water with BM3 as coxswain, SN, SN, and MK3 as bat crew and Gold Team 2 as crew .Gold Team 2 consists of GM3, FN, SN, SN, SN, LTJG, ENS, MK1, SN, and SN. 0721 SHE-1 away to starboard. 0734 SHE-1 alongside to starboard. 0735 SHE-1 raised to the main deck rail. 0740 SHE-1 is ready and cradled for sea with coxswain and boat crew safely on board. 0753 SHE-2 alongside to starboard. 0756 SHE-2 away to starboard. 1224 SHE-2 alongside to port to receive extra boarding team members and water. Boarding team members are SN, BM?, SN, and RD3. 1232 SHE-2 away to port enroute M/V TAMA TIKI. 1238

SHE-2 alongside M/V TAMA TIKI. 1246 Boarding team safely on board M/V TAMA TIKI. 1651 SHE-2 enroute SHERMAN. 1726 SHE-2 away enroute M/V TAMA TIKI after having disembarked 07 boarding team members safely on board. 1740 SHE-2 enroute with boarding team safely on board. 1749 SHE-2 alongside to port. 1753 Boarding team safely on board SHERMAN. 1754 SHE-2 raised to the main deck rail, coxswain and boat crew safely on board.

29 Mar – Underway in the North Arabian Gulf in position 29° 42.8N 049 12.1E conducting MIO OPS. NR1 MDE is on line turning for 3.5 knots. Dog Zebra and material condition Yoke are set throughout the ship with EMCON DELTA in effect. SHERMAN is under the OPCON and ADCON of COMFIFTHFLT. 0640 SHE-2 lowered to the main deck rail. (Part of log unreadable) Boarding Team Blue consists of LTJG, RD2, SK3, ENS, SN, GM3, SN, and SN. 0829 SHE-2 away to port. 0832 SHE-1 lowered to the water with BM1 as coxswain, BM3, MK3, and Blue Team 2 as boat crew. Blue Team 2 consists of MK3, FA, FT2, GM1, SN, and BM3 as boat crew. 0833 SHE-1 away to starboard. 0848 SHE-2 alongside to starboard. 0849 SHE-2 away to starboard. 0854 SHE-1 alongside to starboard. 0856 SHE-1 raised to the rail with coxswain and boat crew safely aboard. 0918 SHE-2 alongside to port. 0919 SHE-2 away to port. 1051 SHE-2 alongside to port. 1053 SHE-2 away to port with Blue Team safely aboard SHERMAN. 1059 SHE-2 alongside to port. 1105 SHE-2 raised to the main deck rail with coxswain and boat crew and Gold Team One safely aboard SHERMAN. 1211 SHE-2 lowered to the water with BM? as coxswain and SN and MK3 as crew. 1220 SHE-2 away to port with 12 Gold Team members enroute to board JORDEN II. Gold Team members consist of LTJG, CWO3, ENS, BM2, SN, SK3, SN, SN, FA, SA, FA, MK?, GM? SN, FT? 1222 SHE-1 lowered to the water with BM3 as coxswain and SN and FN as crew. 1224 SHE-1 away to starboard with 09 POB enroute M/V JORDEN II. 1230 SHE-1 and SHE-2 alongside M/V JORDEN II. 1300 SHE-1 alongside to starboard. 1302 Navy RHI alongside to port, passengers safely on board SHERMAN. 1305 SHE-1 raised and secured to the main deck rail, coxswain and boat crew safely on board. 1318 Navy RHI alongside to port, passengers safely on board, RHI secured to the portside. 1411 Navy RHI away to port enroute M/V AL DBARAN with Blue Team on board. Blue Team consists of LTJG, GM1, ENS, GM3, SN, MK3, BM3, SN, RD2, SK3, SN, SN, FT3. 1419 RHI alongside, Boarding Team Blue safely aboard. 1450 Navy RHI enroute SHERMAN. 1452 RHI alongside to port to receive documents for boarding team. 1511 RHI away to port enroute M/V JORDEN II. 1533 SHE-2 alongside to port. 1535 8 boarding team members safely on board. 1536 SHE-? Away to port. 1538 RHI alongside to port. 1540 Gold Team safely on board. 1541 Navy RHI away to port. 1545 Navy RHI alongside to port and secured on the port side. Boat crew safely on board. 1645 SHE-1 raised and cradled for sea. 1650 SHE-2 alongside to starboard w/01 passengers. 1651 LTJG safely on board 1712 Navy RHI away to port enroute USS PAUL F FOSTER. 1715 SHE-2 alongside to port. 1717 SHE-2 away to port after having safely disembarking 07 boarding team members on SHERMAN. 1720 SHE-2 enroute SHERMAN with Boarding Team Blue safely on board. 1722 SHE-2 alongside to port. 1727 SHE-2 raised to the main deck rail, coxswain and boat crew safely on board. 1804 Navy RHI alongside to starboard. 1806 Navy RHI away to starboard with

09 POB safely on board enroute USS PAUL F FOSTER. <u>1856</u> Secured all standard navigation lights with the exception of the stern light by order of the Commanding Officer.

30 Mar - Underway in the North Arabian Gulf in position 29° 39.5N 049 03.5E conducting MIO OPS. SHERMAN is DIW. NR1 MDE is on line. Dog Zebra and material condition Yoke are set throughout the ship with EMCON DELTA in effect. SHERMAN is under the OPCON and ADCON of COMFIFTHFLT. <u>0420</u> Commenced maneuvering to intercept M/V BINK STAR in position 29° 43.1N 049 06.9E, commenced query of M/V BINK STAR. <u>0814</u> SHE-2 lowered to the water with BM1 (part of log unreadable) lowered to the main deck rail. <u>0831</u> SHE-2 away to port enroute M/V PETRONIA for boarding with 12 POB. <u>0835</u> Boat crew and Boarding Team Gold embarked SHE-1. <u>0837</u> SHE-1 lowered to the water with BM3 as coxswain and MK3 and SN as crewmen. <u>0839</u> SHE-1 away to starboard with 12 POB enroute M/V PETRONIA. Gold Team consists of LTJG, CWO3, ENS, SN, RD3, FT2, FN, SN, BM2, GM3, SN, SN, SN, SN, SN, SN. <u>0844</u> SHE-1 and SHE-2 alongside M/V PETRONIA, Boarding Team Gold safely on board. <u>0855</u> SHE-2 enroute SHERMAN. <u>0856</u> SHE-2 alongside to port. <u>0858</u> SHE-2 raised and secured at the rail, boat crew safely on board, BM3, SN, and MK3. <u>0948</u> CGNR-6596 traversed onto deck w/primary tie downs in place.<u>1005</u> LTJG and CWO3 embarked SHE-1, SHE-1 enroute SHERMAN. <u>1007</u> SHE-1 alongside to starboard, 2 passengers safely aboard. <u>1009</u> SHE-1 away to starboard enroute M/V PETRONIA. <u>1031</u> SHE-2 alongside to starboard. <u>1033</u> 01 passenger AST3 safely aboard SHE-1 <u>1041</u> HS3 safely on board SHE-1. <u>1043</u> SHE-1 away to starboard enroute M/V

PETRONIA. 1750 SHE-1 enroute SHERMAN with 14 POB. 1800 SHE-1 alongside to starboard, Boarding Team Gold safely aboard. 1808 SHE-1 away to starboard enroute M/V PETRONIA to pick up remaining boarding team. 1812 SHE-1 alongside M/V PETRONIA, boarding team on board. 1815 SHE-1 enroute SHERMAN. 1822 SHE-1 alongside to starboard. 1827 SHE-1 cradled for sea. 1828 Boarding team safely on board. 1829 SHE-1 raised to the rail, boat crew safely on board. 1830 SHE-1 cradled and secured for sea. 1843 Boat crew embarked SHE-2, SHE-2 lowered to the waters edge with BM3 as coxswain, SN, and MK3 as crew. 1845 2 PAUL F FOSTER passengers safely on board. 1847 SHE-2 enroute PAUL F FOSTER 5 POB. 1852 SHE-2 alongside PAUL F FOSTER, passengers safely on board. 1854 SHE-2 enroute SHERMAN. 1858 SHE-2 alongside to port. 1900 SHE-2 secured at the rail, boat crew safely on board.

31 Mar - Underway in the North Arabian Gulf in position 29° 43.6N 049 10.6E conducting MIO OPS. NR2 MDE is on line turning for 3.5 knots. Dog Zebra and material condition Yoke are set throughout the ship with EMCON DELTA in effect. SHERMAN is under the OPCON and ADCON of COMFIFTHFLT. 0331 Commenced maneuvering to intercept vessel in position 28° 53N 048 5.1E on course 350° True at 4 knots. 0356 Energized all standard navigation lights. 0740 Maneuvering to intercept M/V IBM HAZM in position 29° 14N 049 22E, vessel is at anchor. 0803 On scene with M/V IBM HAZM in position 29° 12N 049 21E. 0812 SHE-2 Lowered to the water with BM1 as coxswain, SN and MK3 as boat crew. 0822 Boarding Team Blue consisting of LTJG as boarding officer, ENS, GM1, RD2, GM3, SK3, SNBM, SN, and SN as boarding team safely aboard SHE-2. 0824 SHE-2 away to port with 12POB enroute M/V IBM HASM. 0830 SHE-2 alongside M/V IBM HASM. 0833 Remainder of Boarding Team Blue: BM2, FT2, MK3, FN, and SN safely aboard SHE-1. 0834 SHE-1 lowered to the water with BM3 as coxswain, BM?, MK3 and SN as boat crew. 0835 SHE-1 away to starboard with 07 POB enroute M/V IBM HAZM. 0845 Boarding team Blue safely aboard M/V IBM HAZM. 0900 SHE-2 alongside to port. 0902 Coxswain and boat crew safely aboard SHERMAN. 0945 SHE-1 enroute SHERMAN. 0948 SHE-1 alongside to starboard 0954 SN, SN, SN, SN and SA safely aboard SHE-1. 0950 SHE-1 away to starboard enroute M/V IBM HAZM, with 09 POB. 1000 SHE-1 alongside M/V IBM HAZM. 1002 05 members safely aboard. 1019 SHE-1 enroute SHERMAN. 1022 SHE-2 raised to the main deck rail. 1024 Set Flight Con 3 for VERTREP. 1025 SHE-1 alongside to starboard. 1026 SHE-1 away with access kit enroute M/V IBM HAZM. 1035 Desert Duck over the deck. 1036 Commenced VERTREP with Helo. Late Entry – 1030 SHE-1 alongside M/V IBM HAZM. 1050 Secured VERTREP with Helo having transferred mail. 1052 Secured Flight Con 3. 1057 Traversed CGNR-6596 onto the flight deck, primary and secondary tie downs in place. 1122 SHE-1 alongside to starboard. 1152 SHE-1 away to starboard. 1200 SHE-1 alongside M/V IBM HAZM. 1208 SHE-1 away from M/V IBM HAZM enroute PAUL F FOSTER. 1219 SHE-1 alongside PAUL F FOSTER, FSC safely on board. 1224 SHE-1 away enroute SHERMAN. 1223 SHE-1 alongside to starboard. 1224 FSC safely on board SHERMAN, SHE-1 away to starboard enroute M/V IBM HAZM. 1230 SHE-1 alongside M/V IBM HAZM. 1245 Set Flight CON 1 for launch removed secondary tie downs. 1306 Helo off deck and away to port 03

POB. <u>1315</u> Secured from Flight Con 1, set Flight Con 2. <u>1334</u> SHE-1 enroute SHERMAN. <u>1335</u> SHE-1 alongside to starboard. <u>1338</u> SHE-1 away with Captain D.W. Ryan and 02 POB safely on board enroute M/V IBM HAZM. <u>1340</u> SHE-1 alongside M/V IBM HAZM 03 POB safely on board. <u>1423</u> 03 POB safely on board SHE-1. SHE-1 away enroute SHERMAN. <u>1427</u> SHE-1 alongside to starboard Commanding Officer and 02 POB safely on board SHERMAN. <u>1450</u> Set Flight Con 1 for landing, 03 POB <u>1458</u> Helo on deck commenced Hot Refuel with Helo. ???? Secured Hot Refuel with Helo having transferred 118 gallons of JP-5 to Helo. ???? Helo off deck and away to port, 04 POB. <u>1514</u> Secured from Flight Con 1, set Flight Con 2. <u>1620</u> Set Flight Con 1 for landing, 04 POB. <u>1627</u> Helo on deck with TALON engagement. <u>1630</u> Commenced Hot Refuel on deck with Helo. <u>1632</u> Secured Hot Refuel on deck with Helo, transferred no JP-5. <u>1636</u> Secured from Flight Con 1. <u>1712</u> SHE-1 enroute SHERMAN with 10 members of Boarding Team Blue. <u>1715</u> SHE-1 alongside to starboard. <u>1720</u> 10 Boarding team members safely on board SHERMAN. SHE-1 away to starboard enroute M/V IBN HAZM. <u>1725</u> SHE-1 alongside M/V IBN HAZM. <u>1728</u> SHE-1 enroute SHERMAN with 13 POB. <u>1734</u> SHE-1 alongside to starboard. <u>1737</u> All boarding team members safely on board SHERMAN. <u>1742</u> SHE-1 raised to the rail, boat crew safely aboard. <u>1746</u> SHE-1 cradled and secured for sea. <u>1914</u> Secured all standard navigation lights by order of the Commanding Officer.

01 Apr - Underway in the North Arabian Gulf in position 29° 43.8N 049 18.8E conducting MIO OPS. SHERMAN is DIW. Dog Zebra and material condition Yoke are set throughout the ship with EMCON DELTA in effect. SHERMAN is under the

OPCON and ADCON of COMFIFTHFLT. 1540 Traversed Helo onto flight deck primary and secondary tie downs in place. 1620 Set Flight Quarters Condition 1 for launch. 1643 Helo off deck and away to port, 03 POB. 1648 Secured from Flight Quarters Condition 1, set Flight Con 2. 1751 Set Flight Quarters Condition One for landing. 1806 Helo on deck with primary tie downs in place. 1807 Commenced Hot Refuel on deck with Helo. 1810 Secured refuel transferred 80 gallons of JP-5. 1815 CGNR-6596 away to port with 05 POB. 1818 Secured Flight Con 2. 1911 Set Flight Con 1 for landing. (Part of log unreadable) 1934 Commenced Hot Refuel on deck with Helo. 1937 Secured Hot Refuel having transferred 91 gallons of JP-5. 1941 Secured Flight Quarters Condition 1 and red white red white lights. 2135 Traversed Helo into hanger with primary, secondary, and foul weather tie downs in place.

02 Apr - Underway in the North Arabian Gulf in position 29° 42.4N 049 13.0E conducting MIO OPS. SHERMAN is DIW. Dog Zebra and material condition Yoke are set throughout the ship with EMCON DELTA in effect. SHERMAN is under the OPCON and ADCON of COMFIFTHFLT. 0820 Set Flight Quarters Condition one for launch. 0845 Helo is off deck and away to port with 03 POB. 0847 Secured Flight Quarters Condition One set Flight Quarters Condition Two. 1014 Set Flight Quarters Condition One for landing. 1021 Helo is on deck with primary tie downs in place. 1024 Commenced Hot Refuel on deck with Helo. 1030 Secured Hot Refuel on deck after having transferred 55 gallons of JP-5 to Helo. 1031 Secured Flight Quarters Condition One. 1405 SHE-2 lowered to the water with BM2 as coxswain, SN, MK3 AMT2, and LTCDR as boat crew. 1410 SHE-2 away to port enroute USS

PAUL F FOSTER. 1420 SHE-1 lowered to the main deck rail. 1423 SHE-1 lowered to the water with BM1 as coxswain, BM3, SN, SN, and MK3 as boat crew. 1424 SHE-1 away to starboard. 1452 SHE-1 is alongside to starboard. 1454 SHE-1 away to starboard. 1500 SHE-1 alongside to starboard. 1510 Climbing Team safely aboard SHE-1. Climbing Team consists of LTJG and ENS as boarding officers, RD3, AST3, SN, SN, SN (part of log unreadable). 1511 SHE-2 alongside to starboard. 1513 Remaining members of Climbing Team safely aboard SHERMAN. 1519 SHE-2 away to starboard enroute M/V CARIA with 05 POB. 1515 SHE-1 alongside M/V CARIA with Climbing Team safely aboard. 1518 SHE-2 alongside M/V CARIA with Climbing Team safely aboard. 1523 SHE-1 and SHE-2 away to do training. 1715 SHE-1 and SHE-2 alongside. 1759 SHE-2 away to port enroute M/V CARIA 1800 SHE-1 away to port. 1810 SHE-1 and SHE-2 alongside M/V CARIA. 1913 SHE-1 enroute SHERMAN with 11 POB. 1915 SHE-1 alongside to starboard. 1919 SHE-2 alongside to port. 1923 SHE-1 raised to the rail with coxswain boat crew and Climbers safely on board. 1928 SHE-1 cradled and secured for sea. 2203 Set Flight Quarters Condition One for launch. 2228 Commenced Hot Refuel on deck with Helo. 2232 Secured Hot Refuel on deck after having transferred 96 gallons of JP-5 to the Helo. 2230 Helo is off deck and away to port wit 03 POB. Secured Flight Quarters Condition One. 2242 Secured all standard navigation lights by order of the Commanding Officer.

03 Apr - Underway in the North Arabian Gulf in position 29° 42.6N 049 14.1E conducting MIO OPS. SHERMAN is DIW. Dog Zebra and material condition Yoke are set throughout the ship with EMCON DELTA in effect. SHERMAN is under the

OPCON and ADCON of COMFIFTHFLT. 0317 Experienced radar casualty 0330 Set Flight Quarters Condition 1 for landing. 0352 CGNR-6596 on deck 0400 secured from Flight Con 1, primary and secondary tie downs placed on CGNR-6596. 1124 Traversed Helo into hanger primary and secondary tie downs in place. 1130 Set Flight Quarters Condition Three for VERTREP. 1146 Navy Helo 746 over deck. 1147 Commenced VERTREP with Helo. 1149 Secured from VERTREP with Navy Helo away to port. 1150 Secured from Flight Con 3. 1544 SHE-2 lowered to the main deck rail. 1649 Coxswain and boat crew safely aboard SHE-2. 1650 SHE-2 lowered to the water and away to port enroute USS PAUL F FOSTER with BM3, SN, FN, and MK3 as boat crew. BM3 is coxswain. 1703 SHE-2 alongside USS PAUL F FOSTER. 1705 02 passengers safely on board SHE-2. 1705 SHE-2 away enroute SHERMAN. 1707 SHE-2 alongside to port. 1708 02 passengers safely aboard SHERMAN. 1710 SHE-2 raised to the main deck rail. 2150 Traversed Helo onto flight deck with primary and secondary tie downs in place. 2215 Set Flight Quarters Condition One for launch. 2350 Helo off deck and away to port with 03 POB. 2351 Secured Flight Quarters Condition One, set Flight Quarters Condition Two. Secured all standard navigation lights by order of Commanding Officer.

04 Apr - Underway in the North Arabian Gulf in position 29° 30.5N 049 00.1E conducting MIO OPS. Both MDE's are on line turning for 10 knots. Dog Zebra and material condition Yoke are set throughout the ship with EMCON DELTA in effect. SHERMAN is under the OPCON and ADCON of COMFIFTHFLT. 0320 Set Flight Quarters Condition 1 for landing. 0334 CGNR-6596 above deck. 0335 Helo on deck primary tie downs in place. 0340 Secured from Flight Con 1. 0402 CGNR-6596 traversed to hanger, primary, secondary and heavy weather tie downs in place. 0900 Set General Quarters Condition 1 for Gunnery Exercise, structural ZEBRA set throughout the ship. 0940 Commenced CIWIS Gunnery Exercise. 0946 Secured CIWIS Gunnery Exercise having expanded 300 rounds of 25mm with no apparent casualties. 1410 SHE-2 lowered to the water coxswain and boat crew safely aboard. 1411 SHE-2 away with BM2 as coxswain, SNBM and MK3 as boat crew. 1413 SHE-2 alongside to USS PAUL F FOSTER. 1420 SHE-2 away enroute SHERMAN with 03 POB. 1425 SHE-2 alongside to port, 03 passengers safely onboard SHERMAN. 1428 SHE-2 raised to the main deck rail coxswain and boat crew safely aboard 1434 SHE-2 cradled for sea. 1447 Set the UNREP Bill. 1501 Commenced approach on USNS KANAWHA (T-AO-196) 1507 Alongside USNS KANAWHA. 1510 USNS KANAWHA passed messenger. Passed bridge to bridge phone line to USNS KANAWHA. 1515 USNS KANAWHA passed span wire. 1518 Tensioned pan wire. 1520 Received and coupled Fueling Hose. 1523 Commenced Fueling. 1600 Secured fueling after having transferred 54, 195 gallons of JP-5. 1605 Uncoupled and returned Fuel Hose. 1607 All lines are clear. 1613 Secured the UNREP Bill.

05 Apr - Underway in the North Arabian Gulf in position 27° 43.4N 050 49.5E conducting MIO OPS enroute Mina Salman, Bahrain. Both MDE's are on line turning for 15 knots. Dog Zebra and material condition Yoke are set throughout the ship with EMCON DELTA in effect. SHERMAN is under the OPCON and ADCON of COMFIFTHFLT. 0733 Pilot Boat alongside to port. 0734 Pilot safely on board, pilot boat away to port. 0745 Set the Special Sea Detail. 0758 Placed both MDE's in pilothouse control. 0801 Bow Prop lowered and tested, placed in pilothouse control 0815 Moored starboard side to Mina Salman pier at Bahrain. 0818 Secured from Special Sea Detail. 0840 The OOD shifted the watch from the pilothouse to the quarterdeck.

06 Apr – Moored starboard side to Berth 10 Mina Salman, Bahrain with standard mooring lines doubled under OPCON and ADCON of COMFIFTHFLT. Ships status is CHARLIE. Receiving potable water, sewage via barge, and electricity, telephone services via shore ties. All hands are on authorized liberty with the exception of duty section one. U.S. Navy ships are present. 0120 GM3 and FN did not return from Cinderella liberty. 0221 FN returned to ship by base security. (Part of log unreadable) Hands present or accounted for. 0857 Liberty granted to duty section 1 to expire 0000 07 April for E-4 and below and 0845 09 April for E-5 and above. 1527 GM3 checked on board. (rate and name redacted) arrived PCS from TRACEN Petaluma, and (rate and name redacted) arrived PCS from TRACEN Cape May.

07 Apr – Moored starboard side to Berth 10 Mina Salman, Bahrain with standard mooring lines doubled under OPCON and ADCON of COMFIFTHFLT. Ships status is BRAVO. SSDG #1 is providing electrical power. Receiving potable water, sewage via barge, and telephone services via shore ties. All hands are on authorized liberty with the exception of duty section two. U.S. Navy ships are present. 0000 Liberty expired for all E-4 and below. 0845 Liberty expired for duty section 3. Held muster for duty sections 2 and 3. All hands present or accounted for. 0945 Liberty granted to duty section two to expire no later than 0000 for E-4 ad below and 0845 09 April for E-5 and above.

08 Apr – Moored starboard side to Berth 10 Mina Salman, Bahrain with standard mooring lines doubled under OPCON and ADCON of COMFIFTHFLT. Ships status is CHARLIE. Receiving potable water, sewage via barge, and electricity, telephone services via shore ties. All hands are on authorized liberty with the exception of duty section three. U.S. Navy ships are present. 0845 Liberty Expired for duty section 1. Mustered duty section three and one. (Part of log unreadable).

09 Apr – Moored starboard side to Berth 10 Mina Salman, Bahrain with standard mooring lines doubled under OPCON and ADCON of COMFIFTHFLT. Ships status is CHARLIE. Receiving potable water, sewage via barge, and electricity, telephone services via shore ties. All hands are on authorized liberty with the exception of duty section one. U.S. Navy ships are present. 0215 MK2, SN, SN, and FN are in Bahrain Police Custody. 0535 SN, SN, FN, and MK2 returned aboard by base security. 0915 Diving operations commenced on screws. 0927 Liberty granted to duty section one to expire on board 2359 09APR01 for E-4 and below and 0730 10APR01 for all hands. 0925 Secured from diving operations.

10 Apr - Moored starboard side to Berth 10 Mina Salman, Bahrain with standard mooring lines doubled under OPCON and ADCON of COMFIFTHFLT. Ships status is Bravo Six. Receiving potable water, sewage via barge, and electricity, telephone services via shore ties. All hands are on authorized liberty with the exception of duty section one. U.S. Navy ships are present. 0000 Liberty expired for all E-4 and below. 0240 ENS reported the door to ELEX open and unattended. EMO and SSDF Team leader inspected and secured space. 0930 Set the Special Sea Detail. 0956 Embarked Bahrain Port Control Pilot. 0959 Clutched in both MDE's and placed in pilot house control 1005 Underway in the Arabian Gulf. 1020 Secured Special Sea Detail with the exception of the Navigation and Anchor Detail. 1039 Placed both MDE's in engine room control, disembarked Bahrain Port Control Pilot. 1104 Secured Navigation and Anchor Detail. 1340 Traversed Helo on deck with primary and secondary tie downs in place. 1420 Set Flight Quarters Condition One for launch. 1446 Commenced Hot Refuel (log unreadable) 1454 Helo off deck and away to port with 03 POB. 1457 Secured Flight Quarters Condition One, set Flight Quarters Condition Two. 1514 Commenced MK75, 76MM Gunnery Exercise, commenced fire. 1616 Secured from Gunnery Exercise have expended 1 round with now apparent casualties. 1628 Set Flight Quarters Condition One for landing. 1642 Helo is on deck with primary tie downs in place. 1643 Commenced Hot Refuel on deck after having transferred 87 gallons to JP-5 to Helo. 1649 Secured Flight Quarter Condition One. 1656 Commenced MK75, 76 mm Gunnery Exercise. 1657 Commenced Fire. 1658 Secured from Gunnery Exercise having expended 5 rounds from the 76mm with no apparent casualties

11 Apr - Underway in the Arabian Gulf in position 26° 59.9N 052° 15.6E conducting MIO operations. Both MDE's on line turning for 16.5 knots. Material condition Yoke and Dog Zebra are set throughout the ship with EMCON DELTA in effect. SHERMAN is under the OPCON ADCON and TACON of COMFIFTH FLT. 0644 Completed test of JP-5 fuel. Test results SED <1MG/L, H20 <5PPM, FFSL 1. 0755 Traversed Helo onto flight deck with primary and secondary tie downs in place. 0818 Set Flight Quarters Condition One for launch. 0848 Helo off deck and away to port with 05 POB. 0850 Secured Flight Quarters Condition One Set Flight Quarters Condition Two. 1100 Set Flight Quarters Condition One for landing. 1114 Helo is on deck with primary tie downs in place. 1115 Commenced Hot Refuel on deck with Helo. 1119 Secured Hot Refuel on deck after having transferred 133 gallons of JP-5 to the Helo. 1123 Secured Flight Quarters Condition One. 1154 Set Flight Quarters Condition 1 for launch. 1208 Helo off deck and away to port with 03 POB. 1212 Secured Flight Con 1, set Flight Con 2. 1540 Set Flight Quarters Condition One for landing. 1550 Secured from Flight Con 1, set Flight Con 2. 1622 Set Flight Quarters Condition one for landing. 1632 Helo on deck with TALON engagement. 1634 Commenced Hot Refuel on deck with Helo. 1636 Secured Hot Refuel having transferred 73 gallons of JP-5. 1641 Secured from Flight Quarters Condition One. 1818 Traversed Helo into hanger, primary, secondary, and heavy weather tie downs in place.

12 Apr - Underway in the Arabian Gulf in position 27° 03.4N 052° 18.1E conducting MIO operations. NR2 MGT is on line turning for 20 knots. Material condition Yoke and Dog Zebra are set throughout the ship with EMCON DELTA in effect. SHERMAN is under the OPCON ADCON and TACON of COMFIFTH FLT. 0550 SHE-2 lowered to the main deck rail with BM3 as coxswain and BM3 as boat crew 0551 SHE-2 away to port 0558 SHE-2 alongside to port embarked Boarding Team consisting of LTJG as Boarding Officer and MKC (part of log unreadable). 0600 SHE-2 away to port. 0604 SHE-2 alongside M/V ZAINAB. 0606 Boarding Team safely aboard M/V ZAINAB. 0640 Assumed custody of M/V ZAINAB. 0645 Navy RHI alongside to port. 0648 2 personnel safely on board SHERMAN. 0651 Navy RHI away to port. 0700 SHE-2 alongside port. 0705 2 POB safely on board SHE-2 0706 SHE-2 away to port wit 5 POB enroute USS PAUL F FOSTER. 0707 SHE-1 lowered to the 01 deck rail. 0710 Boat crew embarked SHE-1 with BM? as coxswain and SN, SN, and MK3 as crew. 0711 SHE-2 alongside USS PAUL F FOSTER 02 passengers safely aboard. 0712 SHE-2 away PAUL F FOSTER enroute SHERMAN 0713 SHE-1 lowered to the water. 0714 SHE-1 away to starboard 0716 SHE-2 alongside to port. 0718 SHE-2 raised to the rail, boat crew safely embarked SHERMAN. 0719 SHE-2 Secured for sea at the rail. 0721 SHE-1 enroute SHERMAN to pick up radio. 0725 SHE-1 alongside to starboard, received COMCO and handheld GPS. 0726 SHE-1 away to starboard to deliver equipment to boarding team. 0730 SHE-1 alongside M/V ZIANAB. 0735 SHE-1 delivered equipment and will maintain station astern M/V ZAINAB. 0738 Set GQ Condition 1 for PACFIRE. 0754 Commenced 76mm Gunnery Exercise. 0755 Commenced fire at target. 0756 Secured from Gunnery Exercise having expanded 6 rounds of 76mm with no apparent casualties. 0842 SHE-1 lowered to the water. 0843 SHE-1 away to starboard with 05 POB and pump. 0856 SHE-1 alongside M/V ZIANAB, WEPS and equipment on board. 0918 Set Flight Quarters Condition One for landing. 0922 Set Flight CON 3 for VERTREP. 0933 Desert Duck 746 above deck. 0935 Commenced VERTREP with Desert Duck. 0944 Secured from VERTREP having transferred 4 passengers and 4 loads of cargo onto SHERMAN. 0946 Desert Duck away to port. 0949 Desert Duck 746 above deck. 0955 Desert Duck away to port having received their VERTEC'S and transferred one cargo bag. 0957 Secured from Flight Quarters Condition 3. 1034 SHE-1 alongside to starboard with WEPS and LT as passengers. 1038 SHE-1 raised to the rail, boat crew, passengers and cargo safely aboard 1143 SHE-2 enroute M/V ZIANAB with BM2 as coxswain, BM3 and MK3. 1152 Coxswain and boat crew safely on board SHE-1. BM3 as coxswain and MK3 and SN as boat crew. SHE-2 alongside M/V ZIANAB, Boarding safely on board M/V ZIANAB. 1153 SHE-2 away enroute SHERMAN. 1155 SHE-1 lowered to the water and away to starboard enroute M/V ZIANAB with 25 gallons of JP-5. 1157 SHE-2 alongside to port. 1159 Set the Anchor Detail. 1200 SHE-2 raised to the main deck rail. 1202 SHE-1 alongside M/V ZIANAB off loaded 25 gallons of JP-5. 1220 Anchored in position 27° 21.7N 051° 01.2E in 208 feet of water mud bottom. 1224 secured the Anchor Detail. 1228 SHE-2 lowered to the water with BM2 and SN. 1232 04 Divers safely on board SHE-2. 1233 SHE-2 away to port for dive operations. 1234 SHE-1 alongside to starboard. 1235 Commenced Dive Operations off stern. 1236 SHE-1 secured alongside starboard. 1320 SHE-1 away to starboard with LTJG RD3. GM?, MK1, SN, SN, FA enroute M/V ZIANAB. 1355 SHE-1 alongside M/V ZIANAB. 1400 Boarding Team members safely on board M/V ZIANAB. 1432 SHE-2 alongside to port. Secured from Dive Operations. 1436 SHE-2 raised to the main deck rail divers, coxswain and boat crew safely on board 1445 Set the Anchor Detail. 1451 Clutched in both MDE's. 1454 Commenced heaving around on the port anchor. 1500 Anchors aweigh, underway in the Arabian Gulf. 1515 Secured Anchor Detail. 1540 Set Flight Con 3 for Hoist. 1554 Navy Helo 746 over deck. Commenced Hoisting Operations. 1556 Helo away to port. 1605 Hel over deck. 1620

Secured Hoisting Operations after having transport 04 passengers to Helo and cargo. <u>1628</u> Helo away to port. <u>1635</u> Traversed Helo onto Flight deck, primary and secondary tie downs in place. <u>1645</u> Set Flight Con 1 for launch, removed secondary tie downs. <u>1732</u> Secured Flight Quarters Condition 1. <u>1912</u> SHE-1 away to starboard enroute M/V ZIANAB. <u>1932</u> SHE-1 alongside M/V ZIANAB, embarked 04 members of Boarding Team. Members are ENS, RD2, SN, and SN. <u>1934</u> SHE-1 enroute USS CATAWABA. <u>1945</u> SHE-1 alongside USS CATAWABA, 04 members aboard. 1947 SHE-1 enroute SHERMAN. <u>1950</u> Traversed CGNR-6596 into the hanger, primary, secondary, and heavy weather tie downs in place.

Note – Coalition aircraft from USS TRUMAN struck an anti-aircraft site enforcing the Southern No-Fly Zone.

13 Apr - Underway in the Arabian Gulf in position 26° 26.3N 052° 47.3E conducting MIO operations. Both MGT's are on line turning for 27 knots. Material condition Yoke and Dog Zebra are set throughout the ship with EMCON DELTA in effect. SHERMAN is under the OPCON ADCON and TACON of COMFIFTH FLT. Four members of Boarding Team Gold are on board USS CATAWBA. Boarding Team members consist of SN, SN, RD?, and ENS. Remaining members of Boarding Team Gold are on board M/V ZIANAB. Boarding Team members consist of FN, FN, SN, GM3, BM3, MK1, and LTJG. <u>0540</u> SHE-2 lowered to the main deck rail. <u>0546</u> SHE-1 lowered to the main deck rail. <u>0609</u> SHE-2 lowered to the water with SN as coxswain, BM1 and MK3 as boat. <u>0611</u> SHE-2 away to port enroute USS STETHEM. <u>0615</u> SHE-2 alongside USS STETHEM to starboard. <u>0630</u> SHE-1 lowered to the water with BM3 as coxswain, SN and MK3 as boat crew, SHE-2 away to starboard with BM3 and LT as passengers. 0640 SHE-1 away to port with boat crew and MKCM Semler, HSC Beck, ENS, LTJG, and SK3 as passengers. 0645 SHE-1 alongside M/V DIAMOND with 05 passengers safely aboard. 0749 SHE-2 alongside M/V DIAMOND. 0751 SHE-2 away with 05 passenger's enroute SHERMAN. 0758 SHE-2 alongside to port 0800 LTJG, MKCM Semler, HSC Beck, MKC, and LT safely on board SHERMAN. 0801 SHE-2 raised to the main deck rail coxswain and boat crew safely aboard. 0808 SHE-2 cradled. 0855 SHE-2 lowered with boat crew and 03 passengers safely on board SHE-2. 0858 SHE-2 away to port enroute M/V DIAMOND. 0900 SHE-2 alongside M/V DIAMOND 03 passengers, SN, SN, SN safely on board, SHE-2 away enroute SHERMAN. 0903 SHE-2 alongside to port, raised to the main rail. (????) SHE-2 lowered to the water and away to port enroute M/V DIAMOND. SN safely on board M/V DIAMOND, SN safely on board SHE-2. 0922 SHE-2 away enroute SHERMAN. 0924 SHE-2 alongside to port. 0925 SN safely on board SHERMAN. SHE-2 raised to the main deck rail coxswain and boat crew safely on board. 1005 SHE-2 cradled for sea. 1350 Traversed CGNR-6596 onto flight deck, primary tie downs in place. 1410 Set Flight Quarters Condition 1 for launch, removed secondary tie downs from Helo. 1451 Commenced Hot Refuel on deck with Helo. 1453 Secured Hot Refuel having transferred 50 gallons of JP-5 to Helo. 1455 removed primary tie downs from Helo. 1456 Helo is off the deck and away to port with 04 POB. 1453 Secured Flight Con 1, set Flight Con 2. 1630 Set Flight Con 1 for landing. 1639 Helo is on deck with TALON engagement. 1640 disembarked 01 POB. 1641 Commenced Hot Refuel on deck with Helo. 1645 Secured Hot Refuel on deck having transferred 117 gallons of

JP-5 to Helo. 1647 Disengaged TALON. 1650 Helo is off the deck and away to port with 04 POB. Late Entry. 1647 Embarked 01 POB. 1652 Secured Flight Con 1, set Flight Con 2. 1739 SHE-2 lowered to the main deck rail. 1744 SHE-2 lowered to the water with BM3 as coxswain, SN, BM3 and MK3 as boat crew. 1745 SHE-2 away to port. 1749 Observed sunset, energized all standard navigation lights. 1800 SHE-1 alongside to starboard. 1804 SHE-1 away to starboard with BM2 as coxswain SN, MK3 and Boarding Team as boat crew. Boarding Team consists of FT2, SN, RD3, RD3, SN, SN, YNC Planitz, and CWO. 1820 Set Flight Quarters Condition One for landing, energized restricted in ability to maneuver lights. 1842 Helo is on deck with primary tie downs in place. 1845 Commenced Hot Refuel on deck with Helo. 1850 Secured Hot Refuel on deck after having transferred 158 gallons of JP-5 to Helo. 1855 Secured Flight Quarters Condition 1. Restricted maneuver lights secured. SHE-1 alongside to starboard. 1856 SHE-1 raised to the main deck rail with coxswain and boat crew safely aboard. 1912 SHE-1 is ready and cradled for sea. 1916 SHE-2 alongside to port. 1919 SHE-2 raised to the main deck rail with coxswain boat crew and boarding team safely aboard. 1928 SHE-2 cradled and ready for sea. 2016 Traversed Helo into hanger with primary, secondary and foul weather tie downs in place.

14 Apr - Underway in the Arabian Gulf in position 25° 33.1N 055° 05.1E conducting MIO operations. NR1 MDE is on line turning for 8.5 knots. Material condition Yoke and Dog Zebra are set throughout the ship with EMCON DELTA in effect. SHERMAN is under the OPCON ADCON and TACON of COMFIFTH FLT. Four members of Boarding Team Gold are on board USS CATAWBA. Boarding Team members consist of SN, SN, RD?, and ENS. Remaining members of Boarding Team Gold are on board M/V ZIANAB. Boarding Team members consist of FN, FN, SN, RD3, RD3, FT3, YNC Planitz, and CWO. 0700 SHE-2 lowered to the main deck rail. 0702 SHE-2 lowered to the water, HSC Beck, ENS, MKCM Semler, and MK3 all safely aboard. 0704 SHE-2 away to port with SN as coxswain BM? and MK3 as boat crew enroute M/V QAMAR. 0713 SHE-2 alongside M/V QAMAR, disembarked 04 passengers. 0714 04 passengers safely aboard M/V QAMAR, SHE-2 enroute SHERMAN. 0716 SHE-2 alongside to port, BM3, SN, SN, and SN safely aboard SHE-2. 0717 SHE-2 away to port. 0724 SHE-2 alongside M/V QAMAR, 04 members of Security Team safely aboard M/V QAMAR. 0726 CWO, YNC Planitz, FT2, RD3, RD3, SN and SA safely aboard SHE-2. 0728 SHE-2 enroute SHERMAN. 0735 SHE-2 alongside to port. 0740 08 members of Security Team safely aboard SHERMAN. SHE-2 away to port enroute M/V QAMAR. 0746 SHE-2 alongside M/V QAMAR, 08 passengers safely aboard. 0747 HSC Beck and MKCM Semler safely aboard SHE-2, enroute SHERMAN. 0750 SHE-2 alongside to port, HSC Beck, MKCM Semler safely aboard SHERMAN. 0757 SHE-1 raised to the main deck rail, coxswain and boat crew safely aboard SHERMAN. 0815 Received a report from USS CATAWBA of M/V ZAINAB taking on water in position 25° 13N 054° 55E. 0841 SHERMAN on scene with M/V ZAINAB in position 25° 13N 054° 053E, set the Rescue and Assistance Bill on the SHERMAN. R&A Team consists of LT, LTJG, MKC and DC3. (Part of log unreadable) 0900 SHE-2 away to port enroute M/V ZAINAB 0905 SHE-2 alongside the M/V ZAINAB. 0902 SHE-2 enroute SHERMAN. 0912 SHE-2 alongside to port. 0914 LTJG, GM1, FT1 and RD3 safely aboard SHE-2. 0915 SHE-2 way to port enroute M/V ZAINAB. 0919 SHE-2 alongside M/V ZAINAB, 04 members safely aboard M/V ZAINAB. 0922 SHE-2 standing by off the M/V ZAINAB. 0939 SHE-2 enroute SHERMAN. 0946 Traversed CGNR-6596 onto flight deck, primary and secondary tie downs in place. 0941 SHE-2 alongside to starboard to receive life jackets, SHE-2 away enroute M/V ZAINAB. 0949 SHE-2 alongside M/V ZAINAB. 0952 SHE-2 standing off M/V ZAINAB. 1015 USS CATAWBA small boat ALLEY CAT enroute SHERMAN, with ENS. 1018 ALLEY CAT alongside to port, ENS safely aboard. 1019 ALLEY CAY away to port. 1048 SHE-2 alongside M/V ZAINAB, 07 members of Security Team safely aboard SHE-2 1054 SHE-2 alongside to port LT, LTJG, MKC, MK1, BM2, GM3 and SN safely aboard SHERMAN. 1056 SHE-2 raised to the main deck rail, coxswain and boat crew safely aboard SHERMAN. 1234 SHE-2 lowered to the water with BM? as coxswain and MK3 as boat crew. 1244 Set Flight Quarters Condition One for launch. 1246 U.A.E. Coast Guard 655 on scene. 1246 11 crew members of M/V ZAINAB evacuated safely on board SHE-2. 1248 SHE-2 away 1249 SHE-2 transferred 11 POB onto SHE-1 1250 SHE-2 alongside M/V ZAINAB 1257 Helo is off the deck and away to port with 04 POB. 1258 SHE-1 alongside U.A.E. Coast Guard 655 11 crew members safely transferred to 655. 1259 06 Boarding Team members safely on board SHE-2, 06 Boarding Team members remain on board M/V ZAINAB. 1302 Transferred 06 Boarding Team members from SHE-2 to SHE-1. 1304 SHE-1 alongside to port, FT?, RD3, RD3, GM1, SN, SN safely on board SHERMAN. 1306 SHE-1 away enroute M/V ZAINAB. 1310 Boarding Officer reported flooding reduced by

1 foot 1312 Secured Flight Quarters Condition One, set Flight Quarters Condition Two. 1320 Boarding Officer reported increased flooding recommended total evacuation of M/V ZAINAB. 1322 Commanding Officer concurred, ordered evacuation of remaining Boarding Team members. 1323 SHE-2 alongside M/V ZAINAB. 1325 06 remaining Boarding Team members safely on board SHE-2. 1326 SHE-2 away enroute SHERMAN 1327 SHE-2 alongside to port. 1330 LTJG, DC3, FN, LTJG, FN, and MKC Young safely on board SHERMAN. 1340 M/V ZAINAB submerged in 25 meters of water in position 25° 14.9N 054° 51.5E. 1342 Set Flight Quarters Condition One for landing. 1349 Helo on deck with primary tie downs in place. 1353 Secured from Flight Quarters Condition One. 1355 SHE-1 alongside to starboard. 1358 SHE-1 raised to the main deck rail with coxswain and boat crew safely aboard. 1400 SHE-2 alongside to port 1413 SHE-2 away to port enroute UAE-655 to drop off passports of crew of M/V ZAINAB. 1417 SHE-2 alongside UAE-655. 1419 SHE-2 away enroute SHERMAN. 1426 SHE-2 alongside to port. 1523 SHE-2 raised to the man deck rail with boat crew and coxswain safely aboard. 1544 Set Flight Quarters Condition One for launch. 1557 Commenced Hot Refuel on deck with Helo. 1559 Secured Hot Refuel on deck after having transferred 44 gallon of JP-5 to Helo. 1605 Helo off deck and away to port with 04 POB. 1609 Set Flight Quarters Condition Two, secured Flight Quarters Condition One. 1658 Set Flight Quarters Condition One for landing. 1713 Helo on deck with primary tie downs in place. 1714 Commenced Hot Refuel on deck with Helo. 1717 Secured Hot Refuel after having transferred 110 gallons of JP-5 to Helo. 1721 Secured from Flight Quarters Condition One. 1825 Boat crew embarked SHE-2 with BM3 as coxswain, SNBM and

MK3 as crew. 1827 SHE-2 lowered to the water. 1831 Boarding Team safely on board SHE-2. Boarding Team consists of CWO, YNC Planitz, SN, SN, RD3 and RD3. 1832 SHE-2 away to port enroute to M/V DAIMOND. 1850 SHE-2 alongside to M/V DIAMOND. 1907 SHE-2 alongside to port with 06 Boarding Team members. 1904 Boarding Team members safely on board SHERMAN, Boarding Team consists of ENS, MK3, SN, SN, BM2, and SN. 1907 SHE-2 raised to the rail, boat crew safely on board. 1911 SHE-2 cradled and secured for sea. 1912 CGNR-6596 Traversed into hanger and secured with primary, secondary and heavy weather tie down. 1920 SHE-1 cradled and secured for sea.

15 Apr - Underway in the Arabian Gulf in position 25° 17.2N 054° 055.3E conducting MIO operations. NR1 MDE is on line turning for 8.5 knots. Material condition Yoke and Dog Zebra are set throughout the ship with EMCON DELTA in effect. SHERMAN is under the OPCON ADCON and TACON of COMFIFTH FLT. Security Team Q is on M/V DIAMOND with CWO as boarding Officer and YNC Planitz, FT2, SN, SN, RD3 and RD3 as Security Team. 0617 Traversed CGNR-6596 from hanger 0640 Set Flight Quarters Condition 1 for launch, removed primary and secondary tie downs from CGNR-6596. 0656 Helo away to port. 0706 Secured Flight Quarters Condition 1, set Flight Quarters Condition 2. 0729 SHE-2 lowered to the water with BM1 as coxswain, SN, and MK3 and Boarding Team as boat crew. Boarding Team consists of FT?, ???, SN, SN, ENS, SK3, and SN. 0731 SHE-2 away to port enroute M/V DIAMOND. 0742 SHE-2 is alongside to port. 0748 SHE-2 is away to port. 0759 SHE-2 alongside to port SN, RD3, SN, CWO, YNC Planitz, RD3, MKCM Semler, HSC

Beck safely aboard. 0812 SHE-2 away to port 0829 SHE-2 alongside to ort. 0832 SHE-2 raised to the main deck rail with coxswain and boat crew safely aboard. 0843 Set Flight Quarters Condition One for landing. 0850 Helo on deck with TALON engagement 0855 Commenced Hot Refuel on deck with Helo 0900 Secured Hot Refuel on deck after having transferred 144 gallons of JP-5 to Helo. 0908 Secured Flight Quarters Condition One 0944 SHE-2 lowered to the water with BM1 as coxswain SN (part of log unreadable) 1125 SHE-2 alongside to port LT, LT and CDR safely aboard. 1150 SHE-2 away to port enroute UAE CG vessel to deliver drawings 1154 SHE-2 enroute SHERMAN. 1158 SHE-2 alongside to starboard. 1159 EO safely on board SHE-2. 1200 SHE-2 away to starboard enroute UAE CG vessel 1203 EO safely on board UAE vessel. 1221 EO safely on board SHE-2. 1222 SHE-2 enroute SHERMAN. 1225 EO safely on board SHERMAN. 1233 SHE-2 alongside to port 1234 EO on board SHE-2. 1235 SHE-2 away to port enroute UAE CG vessel with 4 POB. 1238 SHE-2 alongside UAE CG vessel with EO safely on board. 1240 SHE-2 is maintaining station off UAE boat. 1305 EO on board SHE-2. 1306 SHE-2 enroute SHERMAN 4 POB. 1310 SHE-2 alongside to starboard, EO safely on board SHERMAN. 1327 SHE-2 alongside to port. 1328 SHE-2 raised to the rail, boat crew safely on board 1330 SHE-2 secured at the rail. 1330 Set Flight Quarters Condition One for Launch. 1332 SHE-2 cradled and secured for sea. 1356 Helo off deck and away to starboard with 4 POB. ???? Set Flight Quarters Condition One for landing. 1525 CGNR-6596 on deck with primary tie downs installed 01 passenger disembarked Helo. 1536 Commenced Hot Refuel on deck with Helo. 1540 Secured from Hot Refuel having passed 132 gallons of JP-5 fuel. 1545 Secured from Flight Quarters.

1610 SHE-2 lowered to the rail. 1617 Boat crew safely on board SHE-2. Boat crew is BM3 as coxswain and SN and MK3 s crew. 1619 SHE-2 Lowered to the water. 1620 SHE-2 away to port enroute M/V DIAMOND. 1622 SHE-2 enroute SHERMAN. 1623 SHE-2 alongside to port 2 passengers on board SHE-2, passengers are MKCM Semler and MK1. SHE-2 away to port enroute M/V DIAMOND. 1624 Both passengers safely on board M/V DIAMOND. 1625 SHE-2 is maintaining station off of M/V DIAMOND. 1638 MKCM Semler and MK1 safely on board SHE-2, SHE-2 enroute SHERMAN. 1639 SHE-2 alongside to port 2 passengers embarked SHERMAN. 1640 Set Flight Quarters Condition One for launch. 1643 SHE-2 raised to the rail, boat crew safely on board SHERMAN. 1645 SHE-2 Secured at the main deck rail. 1655 Commenced Hot Refuel on deck with Helo. 1657 Secured Hot Refuel having passed 29 gallons of JP-5. 1702 Helo off the deck and away to port with 03 POB. 1705 Set Flight Quarters Condition 2. 1719 Boat crew embarked SHE-2, boat crew consists of BM3 s coxswain and SN and MK3 as crew. 1725 Security Team safely on board SHE-2. Security Team consists of ENS and MKCM Semler as boarding officers and BM2, MK3, SN, SN, SN, AND SN. ???? SHE-2 away to port. 1735 Boarding Team safely on board M/V DIAMOND. 1738 SHE-2 alongside to port, raised to the main deck rail. 1743 SHE-2 cradled for sea. 1826 Set Flight Con 1 for landing. 1851 Helo on deck with TALON engagement. 1853 Commenced Hot Refuel on deck with Helo. 1856 Secured Hot Refuel on deck after having transferred 120 gallons of JP-5 TO Helo. 1900 Secured from Flight Quarters Condition 1. 2025 Traversed Helo into hanger, primary, secondary, and heavy weather tie downs in place.

16 Apr - Underway in the Arabian Gulf in position 25° 58.9N 055° 024.7E conducting MIO operations. Both MDE's are on line turning for 10 knots. Material condition Yoke and Dog Zebra are set throughout the ship with EMCON DELTA in effect. SHERMAN is under the OPCON ADCON and TACON of COMFIFTH FLT. Security Team is on M/V DIAMOND with ENS as boarding Officer and MKCM Semler, BM2, MK3, SN, SN, SN, SN, and SN as Security Team. 0002 SHE-2 lowered to the water with BM1 as coxswain, BM3 and MK3 as boat crew. 0004 SHE-2 away to port. 0017 SHE-2 alongside to port. 0012 SHE-2 raised to the main deck rail, coxswain and boat crew safely aboard SHERMAN. 0237 SHE-2 Lowered to 01 deck rail with BM3 as coxswain, BM3 and MK3 as boat crew. 0241 SHE-2 away to port. 0333 SHE-2 alongside to port. 0342 SHE-2 raised to main deck rail, boat crew safely aboard SHERMAN. 0622 SHE-2 cradled for sea. 0917 SHE-2 lowered and secured at the rail. 0931 Boat crew embarked SHE-2, boat crew consists of BM3 as coxswain, SN and MK3 as crew. 0934 SHE-2 away to port. 1038 SHE-2 alongside to port. 1040 SHE-2 raised to the rail boat crew safely on board SHERMAN. 1044 SHE-2 cradled and secured for sea. 1105 Traversed CGNR-6596 onto deck. 1248 Set Flight Quarters Condition One for launch, removed secondary tie downs. 1252 SHE-2 lowered to the main deck rail. 1315 Helo off deck and away to port, 04 POB. 1308 Secured from Flight Con 1 set Flight Con 2. 1447 Set Flight Quarters Condition One for landing 04 POB. 1501 Helo on deck with TALON engagement. 1504 Commenced Hot Refuel with Helo on deck. 1508 Secured Hot Refuel with Helo on deck (part of log unreadable). 1513 Secured from Flight Con 1. 1518 Primary and secondary tie downs installed on Helo. 1602 Coxswain boat crew and passengers safely aboard SHE-2.

1604 SHE-2 away to port with BM? as coxswain BM3 and MK3 as boat crew and CDR as passenger enroute GULF TROUT 1609 SHE-2 alongside tug GULF TROUT CDR safely on board. 1622 CDR safely on board SHE-2, SHE-2 way enroute SHERMAN. 1624 SHE-2 alongside to port. 1626 SHE-2 raised to the main deck rail coxswain, boat crew and CDR safely on board SHERMAN. 1633 Set Flight Quarters Condition One for launch, removed secondary tie downs from Helo. 1649 Commenced Hot Refuel on deck with Helo. 1651 Secured Hot Refuel on deck with Helo having transferred 23 gallons of JP-5. 1655 Helo off deck and away to port, 03 POB. 1658 Secured from Flight Con 1, set Flight Con 2. 1754 SHE-2 lowered to the main deck rail with BM1 as coxswain MK3 and BM3 as boat crew. 1756 SHE-2 away to port enroute M/V DIAMOND. 1759 SHE-2 alongside M/V DIAMOND, 1801 03 passengers safely aboard, SHE-2 enroute SHERMAN. 1806 SHE-2 alongside to port. 1810 Security Team safely aboard SHE-2 with CWO2 as boarding officer, YNC Planitz, RD3, RD3, SN, SN and FN as Security Team. 1811 SHE-2 away to port enroute M/V DIAMOND. 1815 SHE-2 alongside M/V DIAMOND 1818 07 Members of Security Team safely aboard M/V DIAMOND. 1822 04 Members of Security Team safely aboard SHE-2 members are SN, SN, SN, and SN, SHE-2 enroute SHERMAN. 1824 SHE-2 alongside to port, 04 members safely aboard SHERMAN. SHE-2 away to port. 1827 Set Flight Con 1 for landing. 1851 Helo is on deck with TALON engagement. 1852 Commenced Hot Fuel on deck with Helo. 1858 Secured Hot Refuel having transferred 146 gallons of JP-5 to Helo. 1906 Secured from Flight Con 1. 1913 SHE-2 alongside to port ENS, FSC, BM2 and SN safely aboard SHERMAN. 1918 Security Team including ENS, BM2, FT2,

FT1, SK3, SN, SN and FA, safely aboard SHE-2. 1921 SHE-2 away to port enroute USS CATAWBA. 1947 SHE-2 alongside USS CATAWBA, 08 members of Security Team safely aboard. 1949 SHE-2 enroute SHERMAN. 2038 SHE-2 alongside to port. 2040 SHE-2 raised to the main deck rail, coxswain and boat crew safely aboard SHERMAN 2049 SHE-2 cradled and ready for sea. 2056 Traversed CGNR-6596 into hanger, primary and secondary and heavy weather tie downs in place.

17 Apr - Underway in the Arabian Gulf in position 25° 56.6N 055° 029.6E conducting MIO operations. Both MDE's are on line turning for 15 knots. Material condition Yoke and Dog Zebra are set throughout the ship with EMCON DELTA in effect. SHERMAN is under the OPCON ADCON and TACON of COMFIFTH FLT. Security Team is on M/V DIAMOND with CWO as boarding Officer and SN, YNC Planitz, RD3, FN, RD3 and SN as Boarding Team. Boarding Team on USS CATAWBA consists of FN, SN, ENS, BM2, FT2, SN, SK3 and FT1. 0849 SHE-2 lowered to the 01 deck rail. 0930 Statement was issued to the Master of the M/V DIAMOND that his letter will be forwarded to the proper authorities but at this time he is still responsible for the safely of his vessel and crew. 1222 Traversed Helo onto flight deck, primary and secondary tie downs in place. 1240 Set Flight Con 1 for launch, removed secondary tie downs. 1259 Helo is off deck and away to port with 04 POB 1303 Secured Flight Con 1, set Flight Con 2. 1439 Helo is on deck with TALON engagement. 1442 Helo secured engines and disengaged rotors. 1443 Secured FLIGHT Con 1, set the Helo Refueling Detail. 1449 Secured the Helo Refueling Detail having transferred 123 gallons of JP-5 to Helo. 1540 Set Flight Con 1 for launch. 1557 Disengaged TALON 1558 Helo is off the deck and away to starboard with 04 POB. 1600 Secured Flight Con 1, set Flight Con 2. 1728 Set Flight Quarters Condition One for landing. 1738 Helo is on deck with TALON engagement. 1742 Secured from Flight Quarters Condition One, set the Helo Refueling Bill. 1749 Secured the Helo Refueling Bill after having transferred 137 gallons of JP-5 to the Helo. 1925 Traversed Helo into hanger with primary, secondary tie downs in place. 2030 SHE-2 lowered to the main deck rail with SN as coxswain MK3 and MKCM Semler as boat crew. 2036 SHE-2 away to port enroute M/V DIAMOND. 2128 SHE-2 alongside to port Boarding Team safely on board SHE-2. 2130 Boarding Team consists of FN, SN, ENS, BM2, FT2 SN, SK3 and FT1. 2131 SHE-2 alongside to port 08 passengers safely on board SHERMAN. 2145 03 passengers safely on board SHE-2, passengers are MK1, MK3 and MKC. 2153 SHE-2 away to port enroute M/V DIAMOND. 2153 SHE-2 away to port enroute M/V DIAMOND 06 POB. 2205 SHE-2 alongside M/V DIAMOND. 2207 03 passengers safely on board M/V DIAMOND. 2208 SHE-2 enroute SHERMAN with 4 POB. 2216 M/V DIAMOND aweigh anchor enroute Abu Dubai. 2217 SHE-2 alongside to port. 2218 SHE-2 raised to the rail 01 passenger MKCM Semler and boat crew safely aboard SHERMAN. 2220 SHE-2 raised to the rail. 2227 M/V DIAMOND underway. 2241 SHE-2 cradled and ready for sea

18 Apr - Underway in the Arabian Gulf in position 25° 17.0N 054° 043.9E conducting MIO operations. SHERMAN is D.I.W. NR1 MDE is in standby. Material condition Yoke and Dog Zebra are set throughout the ship with EMCON DELTA in effect. SHERMAN is under the OPCON ADCON and TACON of COMFIFTH FLT. Security Team is on M/V DIAMOND with CWO as boarding Officer and SN, YNC Planitz, SN, RD3, FN, RD3, SN, MK2 and MK1. SHERMAN is currently escorting vessel to Abu Dhabi. 0442 M/V DIAMOND commenced leaking oil. 0529 M/V DIAMOND anchored 0535 SHE-2 lowered to the main deck rail with SN as coxswain MK3, MKC and MKCM Semler as boat crew. ???? SHE-2 away to port. ???? MK3, MKC and MKCM Semler safely aboard M/V DIAMOND. 0809 SHE-2 alongside to starboard, MKCM Semler, MKC and MK1 safely aboard SHERMAN. 0817 BM2, SN, SN, SN and SN safely aboard SH-2 0818 SHE-2 away to starboard enroute M/V DIAMOND. 0822 SHE-2 alongside M/V DIAMOND, 05 Security Team members safely aboard, SHE-2 enroute SHERMAN. 0826 SHE-2 alongside to port, RD3, RD3 SN and SN safely aboard SHERMAN. 0829 ENS and SN safely aboard SHE-2 0831 SHE-2 away to port enroute M/V DIAMOND. 0837 SHE-2 alongside M/V DIAMOND, 02 Security Team members safely aboard. 0839 SHE-2 enroute SHERMAN. 0844 SHE-2 alongside to port 0846 CWO2, YNC Planitz and FN safely aboard SHERMAN. 0847 SHE-2 raised to the main deck rail, coxswain and boat crew safely aboard. 0853 SHE-2 raised and cradled for sea. 1058 SHE-2 lowered to the main deck rail. (Part of log unreadable) 1249 SHE-2 alongside to port. 1251 SHE-2 raised to the main deck rail with coxswain and SN safely aboard. 1255 SHE-2 ready and cradled for sea. 1738 Set Flight Quarters Condition 3 for VERTREP. 1756 Navy Helo NR 26 over deck 1756 Commenced VERTREP with Helo. 1935 Secured VERTREP with Helo having received 26 pallets of cargo. 1957 Set the UNREP Bill 2020 Commenced approach to the USNS PECOS. 2026 Alongside USNS PECOS for refueling. 2046 Tensioned the spanwire. 2056 Received and coupled fuel hose.

<u>2057</u> Commenced Fueling <u>2149</u> Secured Fueling having received 89, 313 gallons of JP-5. (Part of log unreadable). <u>2159</u> Secured the UNREP Bill.

19 Apr - Underway in the Arabian Gulf in position 25° 32.0N 054° 057.1E conducting MIO operations. Both MDE's are on line and turning for 15 knots. Material condition Yoke and Dog Zebra are set throughout the ship with EMCON DELTA in effect. SHERMAN is under the OPCON ADCON and TACON of COMFIFTH FLT. Security Team Q is on M/V DIAMOND which consists of ENS and LTJG Archer as boarding officers and MKC Fontenot, MK1 Vandewarker, FT2, BM2, MK3, SN, SN, SN, SN and FN Huettinger. <u>1045</u> Commenced Commanding Officers Non Judicial Proceedings. <u>1157</u> Secured from Commanding Officers Non Judicial Mast Proceedings with the following results: SN, USCG is awarded two weeks extra duty, 2 weeks restriction to the ship, reduction to E-2, which is suspended for violation of Article 92 and 86 UCMJ. <u>1345</u> SHE-2 lowered to the rail. <u>1356</u> Commenced query of vessel in position 25° 01.8N 056° 037.4E on course 160° T identified to be M/V AL ASSSHAAR, under U.A.E flag. <u>1545</u> Set Flight Quarters Condition One for launch. <u>1613</u> Helo off deck and away to port with 04 POB. <u>1615</u> Set Flight Quarters Condition 2. <u>1718</u> Set Flight Quarters Condition One for landing. <u>1733</u> Helo on deck with Talon engagement. <u>1737</u> Secured from Flight Quarters Condition One, set the Helo Refueling Bill. <u>1743</u> Commenced fueling on deck with Helo. <u>1747</u> Secured fueling on deck with Helo after transferring 107 gallons of JP-5. Secured the Helo Refueling Bill. <u>1926</u> Traversed Helo into the hanger. Primary and secondary tie downs in place. <u>2035</u> SHE-2 cradled and ready for sea.

Note - Coalition aircraft from USS TRUMAN struck an early warning radar site enforcing the Southern No-Fly Zone.

20 Apr - Underway in the Gulf of Oman in position 25° 26.3N 056° 035.7E conducting MIO operations. NR2 MDE is on line and turning for 3.5 knots. Material condition Yoke and Dog Zebra are set throughout the ship with EMCON DELTA in effect. SHERMAN is under the OPCON ADCON and TACON of COMFIFTH FLT. Security Team Q is on M/V DIAMOND which consists of ENS and LTJG Archer as boarding officers and MKC Fontenot, MK1 Vandewarker, FT2, BM2, MK3, SN, SN, SN, SN and FN Huettinger. 0740 Commenced query of vessel in position 25° 46.5N 056° 038.5E on course 190° T vessel identified to be M/V HURMOZ under Honduran flag. 0800 JP-5 test complete, SED less than one MG/L, H20 less than 5 PPM, FSH .19. 1311 Traversed Helo out onto flight deck primary and secondary tie downs in place. 1410 Set Flight Quarters Condition One for launch, removed secondary tie downs. 1434 Helo off deck and away to port, 04 POB 1441 Commenced Commanding Officers Non – Judicial proceedings 1610 Secured from Commanding Officers Non – Judicial Proceedings with the following results: FN, USCG is awarded 3 months suspended bust and pending administrative discharge. 1617 Set Flight Quarters Condition One for landing, 04 POB. 1633 Helo on deck with TALON engagement 1636 Secured from Flight Quarters Condition One, set the Helo Refueling Bill. 1642 Secured the Helo Refueling Bill after transferring 144 gallons of JP-5 to Helo. 1856 Traversed Helo into hanger, primary and secondary tie downs in place.

Note – Went through the Strait of Hormuz chasing vessels.

21 Apr - Underway in the Gulf of Oman in position 25° 28.2N 056° 038.0E conducting MIO operations. NR1 MDE is on line and turning for 8.5 knots. Material condition Yoke and Dog Zebra are set throughout the ship with EMCON DELTA in effect. SHERMAN is under the OPCON ADCON and TACON of COMFIFTH FLT. Security Team Q is on M/V DIAMOND which consists of ENS and LTJG Archer as boarding officers and MKC Fontenot, MK1 Vandewarker, FT2, BM2, MK3, SN, SN, SN, SN and FN Huettinger. 0640 Set Flight Quarters Condition One for launch. Late Entry – 0610 Traversed CGNR-6596 from hanger removed secondary tie downs. 0703 Helo above deck and away to port with 04 POB. 0705 Secured Flight Quarters Condition 1, set Flight Con 2. 0730 SHE-2 lowered to the main deck rail. 0741 Coxswain and boat crew safely on board SHE-2. 0743 SHE-2 lowered to the water and away to port with BM3 as coxswain, SN, and MK3 as boat crew. 0748 SHE-2 alongside to port 0750 SHE-2 away to port. 0835 Set Flight Quarters Condition One for landing. 0842 Helo on deck with TALON engagement. 0846 Helo secured engines and disengaged rotors. 0847 Secured Flight Con1 set the Helo Refueling Bill. 0849 SHE-2 alongside to port 0851 SHE-2 raised to the main deck rail coxswain and boar crew safely on board SHERMAN. 0853 Secured the Helo Refueling Bill after having transferred 125 gallons of JP-5 to Helo. 1042 Coxswain and boat crew safely aboard SHE-2. 1044 SHE-2 lowered to the water and away to port with BM3 as coxswain SN and MK3 as boat crew. 1116 SHE-2 alongside to port. 1118 SHE-2 raised to the main deck rail, coxswain and boat crew safely aboard SHERMAN. 1310 Set Flight Con 1 for launch, removed secondary tie downs from Helo. 1336 Helo of the deck and away to port with 04 POB. 1338 Secured Flight Con 1, set

Flight Con 2. 1514 Set Flight Con 1 for landing. 1526 Helo on deck, primary and secondary tie downs in place. 1528 Commenced Hot Refuel on deck with Helo. 1533 Secured Hot Refuel having transferred 142 gallons of JP-5 to Helo. 1537 Helo is off the deck and away to port with 04 POB. 1551 SHE-2 lowered to the water with BM3 s coxswain, SN and MK3 as boat crew. 1552 SHE-2 away to port. 1618 SHE-2 alongside to port. 1621 SHE-2 raised to the main deck rail, coxswain and boat crew safely aboard SHERMAN. 1630 SHE-2 cradled and ready for sea. 1631 Set Flight Con 1 for landing. 1641 Helo is on deck primary tie downs in place. 1649 Secured Flight Con 1, set the Helo Refueling Bill. 1652 Secured the Refueling Bill after having transferred 92 gallons of JP-5 to the Helo. 1849 Traversed Helo into hanger with primary, secondary, and foul weather tie downs in place.

22 Apr - Underway in the Gulf of Oman in position 25° 56.6N 056° 044.7E conducting MIO operations. NR1 MDE is on line and turning for 3.5 knots. Material condition Yoke and Dog Zebra are set throughout the ship with EMCON DELTA in effect. SHERMAN is under the OPCON ADCON and TACON of COMFIFTH FLT. Security Team Q is in port Bahrain with ENS and LTJG Archer as boarding officers and MKC Fontenot, MK1 Vandewarker, FT2, BM2, MK3, SN, SN, SN, SN and FN Huettinger. 0859 SHE-2 lowered to the main deck rail. 0923 Maneuvered to intercept radar contact in position 26° 24.1N 056° 043.6E on course 085° T at 7 knots. 0940 Radar contact crossed into International waters in position 26° 24.2N 056° 045.5E. 0941 SHE-2 lowered to the water with BM1 as coxswain, MK3 and SNBM as boat crew, Boarding Team Black safely aboard SHE-2. Boarding Team Black consists of LTJG as boarding officer, ENS, MKC Young, GM1, RD2, GM3, SN and FN. 0942 SHE-2 away to port enroute M/V GEORGIOUS. 0943 Radar contact in position 26° 23.8N 056° 045.8E identified as M/V GEORGIOUS. 0947 Boarding Officer, LTJG reported that Master of M/V GEORGIOUS threatened to blow up his vessel if we attempted a boarding. 0947 On scene with M/V GEORGIOUS in position 26° 24.2N 056° 046.1E. 0950 Boat crew reports name painted over on stern is GILLIAN EVERARD, LONDON. 0954 Vessel flying Honduran flag. 0955 Small boat reported Master of M/V GEORGIOUS threatening to ram Dhow 500 yards ahead. 1007 Stood down from pursuit of M/V GEORGIOUS in position 26° 25.0N 056° 048.1E, recalled small boat. 1012 alongside to port, Black Team safely aboard SHERMAN. 1015 SHE-2 raised to 01 deck rail, coxswain and boat crew safely aboard. 1028 M/V GEORGIOUS entered Iranian waters in position 26° 23.6N 056° 049.3E, on course 150°T at 6 knots. ???? Traversed Helo onto flight deck, primary and secondary tie downs in place. 1413 Set Flight Quarters Condition One for launch. 1440 Helo is off the deck and away to starboard with 04 POB. 1443 Secured Flight Quarters Condition One, set Flight Quarters Condition Two. 1551 Set Flight Quarters Condition One for landing. 1601 Helo is on deck with TALON engagement 1605 Secured Flight Quarters Condition One, set the Helo Refueling Bill. 1614 Secured the Helo Refueling Bill after having transferred ???? gallons of JP-5 to the Helo. 1802 Traversed CGNR-6596 into hanger primary, secondary tie downs in place. 1940 SHE-2 raised to the 01 deck and secured for sea.

23 Apr - Underway in the Gulf of Oman in position 25° 33.7N 057° 012.7E conducting MIO operations. NR1 MDE is on line and turning for 3.5 knots. Material condition Yoke and Dog Zebra are set throughout the ship with EMCON DELTA in effect. SHERMAN is under the OPCON ADCON and TACON of COMFIFTH FLT. 0945 Traversed Helo onto flight deck with primary and secondary tie downs. 1045 Set Flight Quarters Condition One for launch. 1113 Helo is off the deck and away to port with 05 POB. 1114 Secured Flight Quarters Condition One, set Flight Quarters Condition Two. 1407 Set Flight Quarters Condition One for landing. 1419 CGNR-6596 on deck with 03 POB. 1424 Secured Flight Quarters Condition One, set the Helo Refueling Detail. 1425 Commenced Fueling Helo. 1429 Secured Fueling the Helo after transferred 133 gallons of JP-5, secured the Helo Refueling Bill. 1506 Secured NR1 MDE for oil leak. 1611 Set Flight Quarters Condition One for launch. 1625 Helo of deck and away to starboard with 03 POB. 1627 Set Flight Quarters Condition Two. 1754 Set Flight Quarters Condition One for landing, 03 POB. 1803 SHE-2 cradled for sea. 1804 Helo on deck with primary tie downs. 1805 Commenced Hot Refuel on deck with Helo. 1809 Secured Hot Refuel on deck after transferring 138 gallons of JP-5 to Helo. 1813 Helo off deck and away to port with 03 POB. 1814 Secured from Flight Con One, set Flight Con Two. 1831 Commenced query of M/V OCEAN STAR in position 25° 23.6N 056° 044.8E. 1945 Set Flight Quarters Condition One for landing. 2008 Helo on deck with primary tie downs. 2011 Secured from Flight Con 1, set the Helo Refueling Bill. 2016 Secured the Helo Refueling Bill after transferring 79 gallons of JP-5 to the Helo. 2205 Traversed CGNR-6596 into hanger, primary and secondary tie downs in place. 2214 Secured aft mast light by order of Commanding Officer.

24 Apr - Underway in the Gulf of Oman in position 25° 02.8N 058° 02.0E conducting MIO operations. NR1 MDE is on line and turning for 3.5 knots. Material condition Yoke and Dog Zebra are set throughout the ship with EMCON DELTA in effect. SHERMAN is under the OPCON ADCON and TACON of CT650.6. 0730 Set Flight Quarters Condition One for launch. Set the Helo Refueling Detail. 0749 Helo started engines and engaged rotors. 0753 Commenced Hot Refuel on deck with CGNR-6596. 0756 Secured Hot Refuel having transferred 76 gallons of JP-5 fuel. 0759 Helo off deck and away to port with 04 POB. 0803 Secured Flight Quarters Condition One, set Flight Quarters Condition Two. 0954 Set Flight Quarters Condition 1 for landing. 0944 CGNR-6596 on deck with primary and secondary tie downs in place. 0946 Helo shut down engines and disengaged rotors. 0949 Secured from Flight Con 1, set the Helo Refueling Detail. 1005 Secured Helo Refueling Detail having transferred 140 gallons of JP-5 fuel. 1310 Set Flight Quarters Condition 1 launch. (Part of log unreadable) Set Flight Quarters Condition One for landing, 03 POB. 1516 Helo on deck with primary tie downs in place. 1520 Helo secured engine and disengaged rotors. 1521 Secured Flight Quarters Condition One, set the Helo Refueling Bill. 1522 Commenced Fueling Helo on deck. 1527 Secured from Fueling the Helo on deck after transferring 119 gallons of JP-5, secured the Helo Refueling Bill. 1703 Maneuvered to intercept M/V GEORGIOS in position 24° 27.9N 059° 23.7E on course 208°T at 6 knots. 1712 SHE-2 lowered to the main deck rail. 1732 On scene with M/V GEORGIOS in position 24° 27.2N 059° 22.4E. 1735 SHE-2 lowered to the water with BM1 as coxswain, BM3 and MK3 as boat crew. 1739 Boarding Team Black consists of LTJG as Boarding Officer, ENS, MKC, BM2, RD2, SN, SN, and FN as boarding team, safely aboard SHE-2. 1740 SHE-2 away to port enroute M/V GEORGIOS. 1745 All members of Black Team safely aboard M/V GEORGIOS, SHE-2 away. 1748 SHE-2 Standing off M/V GEORGIOS. 1759 Boarding Officer reports Master is being compliant. 1801 SHE-2 enroute SHERMAN. 1803 Boarding Officer reports M/V GEORGIOS steering system in inoperative, vessel came DIW. 1806 SHE-2 alongside to port, BM1 safely aboard SHERMAN. 1807 LT, MKCM Semler and SN safely aboard SHE-2. 1810 SHE-2 away to port, enroute M/V GEORGIOS. 1812 SHE-2 alongside M/V GEORGIOS 02 members, LT and MKCM Semler safely aboard. 1813 SHE-2 standing off M/V GEORGIOS. 1821 SHE-1 lowered to the main deck rail. 1840 SHE-1 lowered to the water with BM1 as coxswain, BM3 and MK3 as boat crew. 1841 Boarding Team Gulf consisting of CWO2 as Boarding Officer, FT1, FT3, DC3, RD3, SN, SN as boarding team safely aboard SHE-1. 1842 SHE-1 away to starboard enroute M/V GEORGIOS. 1845 SHE-1 alongside M/V GEORGIOS. 1847 Team Golf safely aboard M/V GEORGIOS. 1928 SHE-1 alongside to starboard, CDR safely aboard SHE-1. 1930 SHE-2 away to starboard enroute M/V GEORGIOS. 1933 SHE-1 alongside M/V GEORGIOS, CDR safely aboard. 1946 Black Team safely aboard SHE-2 enroute SHERMAN. 1951 SHE-2 alongside to starboard, Black Team and LT safely aboard. 1953 SHE-2 away. 1955 SHE-2 alongside port. 2000 SHE-2 raised to the main deck rail, coxswain and boat crew safely aboard SHERMAN. 2005 SHE-2 cradled and ready for sea.

Note – SHERMAN left boarding Team Golf on M/V GEORGIOS and sped off to another contact.

25 Apr - Underway in the Gulf of Oman in position 24° 43.5N 060° 39.5E conducting MIO operations. NR1 MGT is on line and turning for 18 knots. Material condition Yoke and Dog Zebra are set throughout the ship with EMCON DELTA in effect. SHERMAN is under the OPCON ADCON and TACON of CT650.6. Boarding Team Golf is on M/V GEORGIOS with CDR as boarding officer, CWO, FT1, SN, RD3, SN, FT3 and DC3 as boarding team in position 24° 25.0N 059° 35.3E. SHE-1 is alongside M/V GRORGIOS with BM1 as coxswain, BM3 and MK3 as boat crew. 0317 SHE-2 lowered to the main deck rail. 0406 SHE-2 lowered to the water with BM3 as coxswain, SNBM and MK3 as boat crew on scene with M/V JOHANGELA in position 24° 45.0N 062° 06.2E. 0410 SHE-2 away to port. 0411 Commenced query of M/V JOHANGELA. LATE ENTRY – 0410 SHE-2 away with Boarding Team Black consisting of LTJG as Boarding Officer, ENS, MKC, RD2, BM2, FN, SNBM and SN as boarding Team. 0504 SHE-2 alongside M/V JOHNAGELO 0505 Boarding Team safely on board, SHE-2 away to port enroute SHERMAN. 0639 On board query of M/V JOHANGELA completed directed to position 24° 30.0N 060°E. Boarding Team Black will stay on board. LATE ENTRY – 0630 Directed M/V GEORGIOS and M/V JOHANGELO to KEZAR under our custody. 0714 SHE-2 alongside to starboard. 0715 LT safely on board. 0727 Boarding Team Kilo safely on board SHE-2. Kilo consists of YNC Planitz, RD3, RD3, SK3, TC2 and SN. SHE-2 away to starboard enroute M/V JOHANGELO. 0740 SHE-2 alongside the M/V JOHANGELO 0743 LT and Security Team Kilo safely on board M/V JOHANGELO. ???? SHE-1 alongside to starboard. 0823 RD2, BM2, MKC, SN and SNBM safely on board SHERMAN. 0832 SHE-2 away enroute M/V

JOHANGELO. 0835 LTJG and LT safely on board SHE-2. SHE-2 away enroute SHERMAN. 0839 SHE-2 alongside to port. 0841 SHE-2 raised to the main deck rail, LT and LTJG and boat crew safely on board SHERMAN. 0848 SHE-2 secured at the 01 deck rail. 1345 Set Flight Quarters Condition 1 for launch, removed secondary tie downs for Helo. 1408 Helo started engines and engaged rotors. 1415 Removed primary tie downs from Helo. 1417 Helo is off the deck and away to port with 04 POB. 1419 Secured Flight Con 1, set Flight Con 2. 1506 SHE-2 lowered to the main deck rail. 1521 SHE-2 lowered to the water with SNBM as coxswain, MK3 and SNBM as boat crew. 1523 SHE-2 away to port enroute M/V KADE JAH. 1525 SHE-2 (part of log unreadable). 1540 Set Flight Con 1 for landing. 1551 Helo is on deck primary tie downs in place. 1553 Helo secured engines and disengaged rotors. 1554 Secured Flight Con 1, set the Helo Refueling Bill. 1559 Secured the Helo Refueling Bill having transferred 125 gallons of JP-5 to the Helo. 1635 Traversed Helo into hanger primary, secondary, and heavy weather tie downs in place. 1713 SHE-1 alongside to starboard. 1718 SHE-1 away to starboard. 1725 SHE-1 alongside to starboard CDR and MKCM Semler safely aboard SHERMAN. 1735 SHE-1 away to starboard with BM1 as coxswain BM3 and MK3 and boarding team as boat crew. Boarding team consists of CWO, MKC, RD2, BM2, BM3, SN and FN enroute M/V GEORGIOS 1742 SHE-1 alongside to starboard 1744 SHE-1 away with HSC Beck aboard enroute M/V GEORGIOS. 1808 SHE-1 alongside to starboard. 1815 SHE-1 raised to the main deck rail with coxswain, boat crew and boarding team safely aboard. Boarding team consists of CWO, SN, RD3, SN, FT3, DC3 and HS3 Gray.

1825 SHE-1 ia ready and cradled for sea. 1850 SHE-2 ready and cradled for sea.

Note – Per the after action report there wasn't a vessel named M/V JOHANGELO, it was in fact the M/V KADA JAH. It was either recorded incorrectly in the logs or the name wasn't correctly identified when queried and boarded.

26 Apr - Underway in the Gulf of Oman in position 24° 51.9N 060° 18.9E conducting MIO operations. NR2 MDE is on line and turning for 5 knots. Material condition Yoke and Dog Zebra are set throughout the ship with EMCON DELTA in effect. SHERMAN is under the OPCON ADCON and TACON of CT650.6. Boarding Team Golf is on M/V GEORGIOS with CWO as boarding officer, RD2, BM3, MKC, BM2, SN, and FN. Boarding Team Kilo is on board M/V KADE JAH, boarding team consists of ENS as boarding officer and FN, YNC Planitz, SN, SK3, RD3 and RD3. 0640 Set Flight Quarters Condition 3 for VERTREP. 0644 SHE-2 Lowered to the water with BM1 as coxswain, BM2 and MK3 as boat crew. 0648 SHE-2 away to port. 0658 Navy Helo 4824 over the deck, commenced VERTREP with HELO. 0702 USNS MOUNT BAKER (T-AE-34) small boat alongside to port, LTJG Archer, ENS, FT2, SN, SN, SN, SN, FN Huettinger, MKC Fontenot, MK1 Vandewarker safely aboard SHERMAN. 0709 Navy small boat away to port. 0712 Secured VERTREP with Helo having transferred 3 triwalls of mail. 0713 Helo away to port. 0715 Secured Flight Con 3. 0721 SHE-2 alongside to port. 0724 SHE-2 away to port enroute USNS Mount BAKER. 0725 USNS MOUNT BAKER small boat alongside to port. 0728 MOUNT BAKER small boat away to port. 0730 SHE-2 enroute SHERMAN. 0735 SHE-2 alongside to port. 0737 SHE-2 raised to the main deck rail, coxswain and boat crew safely aboard. 0748 SHE-2 cradled and ready for sea. 1341 Set Flight Quarters Condition 1 for launch. (Part of log unreadable) 1410 Secured Flight Quarters Condition One, set Flight Quarters Condition Two. 1430 SHE-2 lowered to the water with SN as coxswain SN and MK3 as boat crew. 1431 CWO, FT1, SN, MK1, RD3, safely aboard SHE-2. 1437 SHE-2 away to port enroute M/V KADA JAH. 1554 SHE-2 alongside to port 1457 ENS, YNC Planitz, FN, SN, SK3, RD3and RD3 safely aboard SHERMAN. 1501 ENS, FT2, MK2, SN, SN and SN safely aboard SHE-2. 1502 SHE-2 away to port enroute M/V GEORGIOS. 1525 Set Flight Quarters Condition One for landing. 1535 Helo on deck with primary tie downs in place. 1539 Helo stopped engines and disengaged rotors. 1540 Secured Flight Quarters Condition One, set the Helo Refueling Bill. 1545 SHE-2 alongside to port. 1547 Secured the Helo Refueling Bill after having transferred 132 gallons of JP-5 to the Helo. 1548 RD2, BM3, MKC, CWO, BM2, SN, FN safely aboard SHERMAN. 1549 SN safely on SHE-2 away to port enroute M/V GEORGIOS. 1606 SHE-2 alongside to port 1609 SHE-2 away to port. 1645 SHE-2 alongside to port 1648 SHE-2 raised to the main deck rail with coxswain boat crew and SN safely aboard SHERMAN. 1650 SHE-2 ready and cradled for sea.

27 Apr - Underway in the Gulf of Oman in position 25° 18.6N 052° 49.6E conducting MIO operations. NR2 MDE is on line and turning for 5 knots. Material condition Yoke and Dog Zebra are set throughout the ship with EMCON DELTA in effect. SHERMAN is under the OPCON ADCON and TACON of COMFIFTHFLT. Boarding Team Golf is on M/V GEORGIOS with ENS as boarding officer, FT2, MK2, SN, SN, SN.

Boarding Team Kilo is on board M/V KADE JAH, boarding team consists of CWO as boarding officer MK1, FT1, RD3, FN and SN. 0104 Visibility decreased to 4,000 yards, energized sound signal for vessel underway, making way. 0149 Visibility restored to 10,000 yards, secured signals. 0318 Visibility decreased to 50 yards, energized sound signal for vessel underway, making way. 0400 Visibility restored to 2,000 yards, secured sound signals. 0718 SHE-2 lowered to the main deck rail 0720 SHE-2 away to port with SN as coxswain, SN and MK3 as boat crew. 0740 SHE-2 alongside to port. 0745 SHE-2 raised to the main rail with coxswain and boat crew safely aboard 0825 Traversed Helo onto Flight deck with primary and secondary tie downs in place. 0840 Set Flight Quarters Condition One for launch, removed secondary tie downs from Helo. 0854 Helo started engines and engaged rotors. 0903 Removed primary tie downs. 0904 Helo off deck and away to port with 04 POB. 0906 Secured Flight Quarters Condition One, set Flight Quarters Condition Two. 1028 Set Flight Quarters Condition One for landing. 1038 Helo is on deck with primary tie downs in place. 1041 Helo stopped engines and disengaged rotors. 1043 Secured Flight Quarters Condition One, set the Helo Refueling Bill. LATE ENTRY – M/V KADE JAH diverted by CF-5. 1048 Secured the Helo Refueling Bill after having transferred 124 gallons of JP-5 to the Helo. 1329 Boat crew embarked SHE-2 with SNBM as coxswain, SN ad MK3 as crew. 1330 SHE-2 lowered to the water 1331 Boarding Team Kilo safely aboard SHE-2. Boarding Team Kilo consists of ENS, SK3, FN, RD3, RD3, FN, SN. 1335 SHE-2 away to port enroute M/V KADE JAH. 1339 SHE-2 alongside to starboard M/V KADE JAH. 1343 CGNR-6596 traversed to the hanger, secured with primary, secondary and heavy weather tie downs. 1357 SHE-2 alongside to port Boarding Team Kilo safely on board SHERMAN. 1400 SHE-2 raised to the 01 deck rail. 1410 SHE-2 lowered to the water. 1412 Boarding Team Gulf safely on board SHE-2. Boarding Team Gulf consists of CWO3, DC3, MKC, BM3, SN, BM3, RD?, BM?. 1414 SHE-2 away to port with 11 POB, enroute M/V GEORGIOS. 1416 SHE-2 away to starboard enroute SHERMAN. 1434 SHE-2 alongside to port boarding team safely on board SHERMAN. 1444 SHE-2 away to port enroute M/V KADE JAH. 1453 Supplies safely on board M/V KADE JAH, SHE-2 returned. 1831 Commenced Commanding Officers Non-Judicial Proceedings. 1834 Crossed into OMANI Territorial Seas in position 26° 18.8N 056° 45.0E. 1947 Secured from Commanding Officers Non-Judicial Proceedings with the following results, FN, USCG was awarded six months suspended bust pending administrative discharge, SA, USCG was awarded 14 days extra duty, SN, USCG was awarded six months bust pending administrative discharge, MK2, USCG was awarded no punishment.

28 Apr - Underway in the Strait of Hormuz in position 26° 36.6N 056° 27.0E conducting MIO operations. NR2 MDE is on line and turning for 5 knots. Material condition Yoke and Dog Zebra are set throughout the ship with EMCON DELTA in effect. SHERMAN is under the OPCON ADCON and TACON of COMFIFTHFLT. Boarding Team Golf is on M/V GEORGIOS with CWO3 as boarding officer, MKC, BM2, RD2, DC3, BM?, BM3 and SN. Boarding Team Kilo is on board M/V KADE JAH, boarding team consists of ENS as boarding officer and RD3, RD3, SK3, SN, FA and FA. (Part of log unreadable). 0704 SHE-2 away to port. 0713 Navy RHI alongside to starboard. 0714 03 passengers safely on board SHERMAN. 0715 Navy RHI away to starboard. 0805 SHE-2 alongside to port to unload equipment. 0806 SHE-2 raised to the rail. 0808 SHE-2 lowered to the water and away to port. 0812 Navy RHI alongside to port. 0813 All passengers safely on board RHI. 0814 Navy RHI away to port. 0822 Boarding Team Kilo handed custody of M/V KADE JAH over to the USS MITSCHER (DDG-57). 0826 Fuel test complete: SED less than one, FW less than five, FSII .17 1833 SHE-2 alongside to port. 0835 Kilo Team safely on board SHERMAN. 0836 SHE-2 away to port enroute M/V GEORGIOS to pick up Team Golf. 0837 Boarding Team Golf handed custody of M/V GEORGIOS over to USS MITSCHER. 0843 SHE-2 alongside to port. 0847 Boarding Team Golf safely on board SHERMAN. 0848 SHE-2 raised to the rail, boat crew and coxswain safely on board 0852 SHE-2 cradled and secured for sea. 0955 Crossed Claimed Iranian Territorial Seas in position 26° 22.7N 055° 33.5E. 1745 Set the UNREP Bill. 1828 Commenced approach on USNS PECOS (T-AO-197). 1829 Alongside the USNS PECOS for refueling. 1832 Passed the bridge to bride phone line to USNS PECOS. 1833 USNS PECOS passed spanwire. 1849 tensioned spanwire. 1851 Received and coupled fuel hose. 1856 Commenced refueling 2000 Completed fueling having received 92,600 gallons of JP-5 fuel. 2002 Uncoupled fuel hose and passed to USNS PECOS. 2003 Detensioned spanwire and passed spanwire to USNS PECOS. 2010 Secured the UNREP Bill.

Note – Coalition aircraft including those from USS TRUMAN targeted Iraqi air defense radar and artillery.

29 Apr - Underway in the Arabian Gulf in position 26° 50.0N 051° 50.8E conducting MIO operations. NR2 MDE is on line and turning for 12.5 knots. Material condition Yoke and Dog Zebra are set throughout the ship with EMCON DELTA in effect. SHERMAN is under the OPCON and ADCON of COMFIFTHFLT and TACON of CTC 50 C. 0825 Set the Navigation and Anchor Detail. 1049 Pilot boat alongside to port. 1050 Pilot safely aboard. 1051 Pilot boat away to port. 1143 Moored starboard side to Berth Five Mina Salman, Bahrain. 1146 Secured Special Sea Detail. 1202 Shifted the watch from the pilot house to the quarterdeck. 1203 SA reported PCS from TRACEN Cape May. FN reported PCS from CGC BLACKTIP. 1229 Liberty granted to duty section One and Two to expire on board no later than 1700 29 April 2001. 1300 BM3 departed PCS to USCGC MIDGETT. RD3 departed TAD to ISC Alameda, CA. FN departed TAD to ISC Alameda. FN departed PCS to TRACEN Yorktown, VA. SN departed TAD to ISC Alameda, CA. FN departed TAD to ISC Alameda, CA. SN departed PCS to TRACEN Yorktown, VA. 1700 Liberty expires for all hands. 1815 Clutched in both MDE's and placed in pilot house control. 1810 Underway in the Arabian Gulf. 1855 Secured the Navigation and Anchor Detail.

30 Apr - Underway in the Arabian Gulf in position 26° 50.2N 051° 22.1E conducting MIO operations. NR1 MDE is on line and turning for 12.5 knots. Material condition Yoke and Dog Zebra are set throughout the ship with EMCON DELTA in effect. SHERMAN is under the OPCON and ADCON of COMFIFTHFLT and TACON of CTC 50 C. 1356 Entered Iranian Territorial waters. 1530 Commenced Commanding Officers Non-Judicial Proceedings under Article 15 UCMJ.

1634 Secured from Commanding Officers Non-Judicial Proceedings with the following results: GM3, USCG is awarded 14 days extra duty for violation of article 92 UCMJ. 1905 Exited Iranian Territorial Seas.

Note – Transited the Strait of Hormuz on our way out of the Persian Gulf.

01 May - Underway in the Gulf of Oman in position 26° 28.3N 056° 36.9E. NR2 MDE is on line and turning for 12.5 knots. Material condition Yoke and Dog Zebra are set throughout the ship with EMCON DELTA in effect. SHERMAN is under the OPCON and ADCON of COMFIFTHFLT and TACON of CTC 50 C. 1252 Traversed CGNR-6596 onto deck. 1357 Set Flight Quarters Condition One for launch. 1410 Helo started engines and engaged rotors. 1441 Helo off deck and away to port with 04 POB. 1444 Set Flight Quarters Condition 2. 1556 Set Flight Quarters Condition One for landing. 1611 Helo on deck with primary tie downs in place. 1614 Helo shut down engines and disengaged rotors. 1615 Secured from Flight Con 1. 1843 Traversed Helo into hanger with primary, secondary and foul weather tie downs in place.

02 May - Underway in the Arabian Sea in position 21° 52.1N 060° 28.4E enroute Seychelles NR1 MDE is on line and turning for 12.5 knots. Material condition Yoke and Dog Zebra are set throughout the ship with EMCON DELTA in effect. SHERMAN is under the OPCON and TACON of COMFIFTHFLT.

03 May - Underway in the Arabian Sea in position 15° 06.7N 059° 25.0E enroute Victoria, Seychelles. Both MDE's are on line and turning for 16 knots. Material condition Yoke and Dog Zebra are set throughout the ship with EMCON DELTA in effect. SHERMAN is under the OPCON and TACON of COMFIFTHFLT. 1510 Traversed Helo with primary and secondary tie downs. 1635 Traversed Helo with primary and secondary tie downs in place.

04 May - Underway in the Arabian Sea in position 09° 32.2N 058° 13.9E enroute Victoria, Seychelles. Both MDE's are on line and turning for 16 knots. Material condition Yoke and Dog Zebra are set throughout the ship with EMCON DELTA in effect. SHERMAN is under the OPCON and TACON of COMFIFTHFLT. 1730 Sest Flight Quarters Condition 1 for launch. 1745 SHE-2 lowered to the main deck rail with BM3 as coxswain, SN, BM3 and MK3 as boat crew. CGNR-6596 started engines and engaged rotors. 1754 Helo above deck and away to port with 04 POB. 1757 SHE-2 lowered to water and away to port with SN as passenger. Secured from Flight Con 1, set Flight Con 2. 1843 Set Flight Con 1 for landing. 1900 CGNR-6596 on deck primary tie downs in place, passenger clear of Helo. 1901 Passed and energized pump for Hot Refuel. 1904 Pumps secured after transferring 120 gallons of JP-5 to Helo. Embarked 04 POB 1908 Helo above deck and away to port. 1911 Secured from Flight Con 1, set Flight Con 2. 2026 Set Flight Con 1 for landing. 2040 CGNR-6596 above deck. 2042 Helo on deck. 2043 passed over pump, energized pump. 2044 Commenced Hot Refuel on deck with Helo. 2046 Secured pump after transferring 113 gallons of JP-5. 2047 Disembarked 01 POB, embarked 01 POB. 2050 Embarked 01 POB. 2055 Helo above deck and away to starboard. 2058 Secured from Flight Con 1, set Flight Con 2. 2139 Set Flight Quarters

Condition One for landing. 2153 Helo commenced touch and go's. 2158 Helo is on deck after completing 02 touch and go's. 2203 Helo stopped engines and disengaged rotors. 2209 SHE-2 alongside to port. 2212 SHE-2 raised to the man deck rail with coxswain and boat crew safely aboard. 2214 Set the Helo Refueling Bill. 2219 Secured the Helo Refueling Bill after having transferred 66 gallons of JP-5. 2221 SHE-2 is ready and cradled for sea.

05 May - Underway in the Arabian Sea in position 03° 21.9N 056° 56.9E enroute Victoria, Seychelles. Both MDE's are on line and turning for 16 knots. Material condition Yoke and Dog Zebra are set throughout the ship with EMCON DELTA in effect. SHERMAN is under the OPCON and TACON of COMFIFTHFLT. 0730 Commenced preparations for entering port in Victoria, Seychelles. 0923 Commenced 25mm and .50 CAL Gunnery Exercise. 0940 Secured 25mm and .50CAL Gunnery Exercise having expanded 800 rounds of .50 CAL and 220 rounds of 25mm ammo with no apparent casualties. 0958 Commenced CIWIS Gunnery Exercise. 1014 Secured CIWIS Gunnery Exercise having expanded 1500 rounds of 20mm CIWIS ammo with no apparent casualties. 1317 SHE-2 lowered to the Main deck rail with BM3 as coxswain, MK3, GM2, and BM3 as boat crew. 1318 SHE-2 lowered to the water and away to port. 1325 Crossed Equator at Longitude 056° 15.0E 132.6. 1530 Declutched both MDE's placed in immediate standby. 1330 Commenced Swim Call. 1400 Secured Swim Call. 1414 Clutched in NR1 MGT. 1416 SHE-2 raised to the main deck rail boat crew safely aboard SHERMAN. 1418 SHE-2 Cradled for sea.

06 May - Underway in the Indian Ocean in position 01° 43.6S 055° 39.4E enroute Victoria, Seychelles. Both MDE's are on line and turning for 16 knots. Material condition Yoke and Dog Zebra are set throughout the ship with EMCON DELTA in effect. SHERMAN is under the OPCON and TACON of COMFIFTHFLT. 0420 Visibility decreased to 4,000 yards, commenced sounding fog signals for a power driven vessel underway making way, set structural ZEBRA main deck and below. 0626 Secured Fog Signal visibility 12,000 yards. 0651 Set the Navigation and Anchor Detail. 0653 Visibility reduced to 50 yards, energized Fog Signal. 0700 Visibility restored to 4,000 yards secured Fog Signal. 0732 Victoria Port Control Pilot alongside to starboard. 0736 Embarked Pilot. 0803 Moored starboard side to Mahe Quay. 0818 Secured Special Sea Detail with exception of Brow Detail and line handlers. Disembarked Pilot. 0837 CWO assumed the inport OOD and shifted the watch from the Pilothouse to the Quarterdeck. 1013 Liberty granted to all hands with the exception of Duty Section 1 to expire onboard 0845 07 May 2001 for Duty Section 2, to expire 0845 08 May 2001 for Duty Section 3, to expire 0845 09 May 2001 for Duty Section 4, and 0830 10 May 2001 for all hands.

07 May – Moored starboard side to Commercial Pier, Victoria, Seychelles with standard mooring lines doubled under the OPCON and ADCON of COMFIFTHFLT. Ships status is Bravo-6. Material condition Yoke is set throughout the ship. All deck, anchor, and aircraft warning lights are energized and burning brightly. NR1 SSDG is providing electrical power. Receiving potable water and telephone services via shore tie. All hands are on authorized liberty with the exception of Duty

Section One. 0845 Liberty expired for Duty Section 2. 0924 Liberty Granted to Duty Section 1 to expire no later than 0830 10 May 2001.

08 May – Moored starboard side to Commercial Pier, Victoria, Seychelles with standard mooring lines doubled under the OPCON and ADCON of COMFIFTHFLT. Ships status is Bravo-6. Material condition Yoke is set throughout the ship. All deck, anchor, and aircraft warning lights are energized and burning brightly. NR1 SSDG is providing electrical power. Receiving potable water and telephone services via shore tie. All hands are on authorized liberty with the exception of Duty Section Two. 0845 Liberty expired for Duty Section 3, held morning muster for Duty Sections Two and Three. All hands present or accounted for with the exception of SN 0859 Liberty granted to Duty Section Two to expire 0830 10 May 2001. 0908 SN is accounted for. 1347 SHE-2 lowered to the water. Boar crew SN, SN, SN, BM3 as coxswain. 1435 SHE-2 cradled for sea. 1452 SA taken to local hospital, XO is present in the hospital.

09 May – Moored starboard side to Commercial Pier, Victoria, Seychelles with standard mooring lines doubled under the OPCON and ADCON of COMFIFTHFLT. Ships status is Bravo-6. Material condition Yoke is set throughout the ship. All deck, anchor, and aircraft warning lights are energized and burning brightly. NR1 SSDG is providing electrical power. Receiving potable water and telephone services via shore tie. All hands are on authorized liberty with the exception of Duty Section Three. 0845 Liberty expired for Duty Section 4, held morning muster for Duty Sections Three and Four. All hands present or accounted for. 0902 Liberty granted to Duty Section Three to expire 0830 10 May 2001.

10 May – Moored starboard side to Commercial Pier, Victoria, Seychelles with standard mooring lines doubled under the OPCON and ADCON of COMFIFTHFLT. Ships status is Bravo-6. Material condition Yoke is set throughout the ship. All deck, anchor, and aircraft warning lights are energized and burning brightly. NR1 SSDG is providing electrical power. Receiving potable water and telephone services via shore tie. All hands are on authorized liberty with the exception of Duty Section Four. 0830 Liberty expired for all hands. 0854 Singled up all mooring lines. 0930 Set the Special Sea Detail. 1007 Underway from Commercial Pier Victoria, Seychelles. 1032 Pilot disembarked, Pilot Boat away to port. 1033 Secured the Special Sea Detail with the exception of the Navigation and Anchor Detail. 1050 Set Flight Quarters Condition One for launch. 1051 Secured Navigation Detail. 1105 CGNR-6596 started engines and engaged rotors. 1120 Helo off deck and away to port. 1123 Set Flight Quarters Condition Two. 1206 Set Flight Quarters Condition 1 for landing. 1220 Helo on deck, primary tie downs in place. 1221 Energized pump for Hot Refueling on deck. 1223 Secured pump after transferring 65 gallons of JP-5 to CGNR-6596. 1224 Embarked 05 POB aboard Helo. 1226 Removed primary tie downs from Helo. 1228 Helo above deck and clear to port. 1255 CGNR-6596 on deck primary tie downs in place. 1257 Disembarked 02 POB 1302 Energized pump for Hot Refuek on deck 1305 Secured pump after transferring 50 gallons of JP-5 to Helo. 1306 Embarked 02 passengers aboard Helo, removed primary tie downs from Helo. 1307 CGNR-6596 above deck and away to port, secured Flight

Con 1, set Flight Con 2. 1335 Set Flight Con 1 for landing. 1342 CGNR-6596 on deck. 1343 Placed primary tie downs in place. 1344 Helo shut down engines and disengaged rotors. 1345 Secured Flight Con 1. 1506 Commenced General Quarters Drill. 1550 Secured General Quarters Drill. 1618 Traversed Helo into hanger with primary and secondary tie downs in place.

11 May - Underway in the Indian Ocean in position 07° 33.9S 055° 37.9E Both MDE's are on line and turning for 15 knots. Material condition Yoke and Dog Zebra are set throughout the ship with EMCON DELTA in effect. SHERMAN is under the OPCON and TACON of COMFIFTHFLT. 1305 Commenced Man Overboard Drill. 1339 Secured from Man Overboard Drill. 1446 Commenced Abandon Ship Drill. 1506 Secured from Abandon Ship Drill.

12 May - Underway in the Indian Ocean in position 012° 58.1S 056° 10.6E enroute Port Louis, Mauritius. NR1 MDE is on line and turning for 12.5 knots. NR2 MDE is OOC due to loss of engine speed control. Material condition Yoke and Dog Zebra are set throughout the ship with EMCON DELTA in effect. SHERMAN is under the OPCON and TACON of COMPACAREA Alameda, CA. 1500 CGNR-6596 traversed onto deck. 1640 Set Flight Con 1. 1650 Helo started engines and engage rotors. 1707 Helo above deck and away to port. 1708 Set Flight Con 4 for HIFIR 1712 CGNR-6596 above deck. 1714 Passed over hose to Helo, energized pumps. 1720 Secured pumps after transferring 117 gallons of JP-5 to Helo. 1721 Passed over hose. 1722 Helo away to port. 1750 Helo on deck, commenced touch and go landings. 1808 Helo above deck and away to port. 1812 Helo above deck, passed over basket. 1813

Basket on deck. Helo clear to port. 1814 Helo passed over basket, basket on deck, Helo clear to port. 1816 Set Flight Con 4 for HIFIR. 1817 Passed over hose. 1821 Energized pump. 1823 Secured pump after transferring 41 gallons of JP-5 to Helo. (Part of log unreadable). Set Flight Con 1, commenced touch and go landings. 1854 Helo on deck after completing 07 touch and go landings. 1855 Helo secured engines and shut down rotors. 1900 Secured Flight Con 1. 1923 Commenced Commanding Officers Non-Judicial proceedings. 1940 Secured from Commanding Officers Non-Judicial proceedings with the following results. SA and GM3 each awarded 50 dollar fine and six months suspension. 2023 Helo hangered with primary and secondary tie downs in place. 2040 Commenced Commanding Officers Non-Judicial proceedings. 2116 Secured from Commanding Officers Non-Judicial proceedings, all charges dismissed.

13 May - Underway in the Indian Ocean in position 018° 20.4S 057° 07.2E enroute Port Louis, Mauritius. Both MDE's are on line and turning for 16 knots. Material condition Yoke and Dog Zebra are set throughout the ship with EMCON DELTA in effect. SHERMAN is under the OPCON and TACON of COMPACAREA Alameda, CA. (part of log unreadable) Port Louis Pilot. 0815 Moored starboard side berth 1 Port Louis Mauritius. 0830 LTJG assumed the inport OOD and shifted the watch from the Pilothouse to the Quarterdeck. 1445 BMC and ET2 arrived from PACAREA TACLET.

14 May – Moored starboard side to berth1 Port Louis, Mauritius with standard mooring lines doubled under the OPCON and ADCON of COMPACAREA. Ships status is Bravo-6. All deck, anchor, and aircraft warning lights are energized and burning brightly. Material Condition Yoke is set throughout the ship. Receiving potable water and sewage via barge and phone lines and electrical service via shore ties. All hands are on authorized liberty with the exception of Duty Section One. Other ships present are commercial fishing vessels. <u>0805</u> Commenced fueling. <u>0845</u> Liberty expired for Duty Section Two. Held muster for Duty Sections One and Two, all hands present or accounted for. <u>0949</u> Completed fueling having taken on 26,694 gallons on F-76, drafts are FWD 14' 6" AFT 15' 0"

15 May – Moored starboard side to berth1 Port Louis, Mauritius with standard mooring lines doubled under the OPCON and ADCON of COMPACAREA. Ships status is Bravo-6. All deck, anchor, and aircraft warning lights are energized and burning brightly. Material Condition Yoke is set throughout the ship. Receiving potable water and sewage via barge and phone lines and electrical service via shore ties. All hands are on authorized liberty with the exception of Duty Section Two. Other ships present are commercial fishing vessels. <u>0845</u> Liberty expired for Duty Section Three, held muster for Duty Sections Two and Three with all hands present or accounted for. <u>0848</u> Water barge departed. <u>0855</u> Liberty granted for Duty Section Two to expire for all hands 17 May 2001 0630.

16 May – Moored starboard side to berth1 Port Louis, Mauritius with standard mooring lines doubled under the OPCON and ADCON of COMPACAREA. Ships status is Bravo. All deck, anchor, and aircraft warning lights are energized and burning brightly. Material Condition Yoke is set throughout the ship. Receiving potable water and sewage via barge and phone lines and electrical service via shore ties. All hands are on authorized liberty with the exception of Duty Section Three. Other ships present are commercial fishing vessels. 0845 Liberty expired for Duty Section Four, held muster for Duty Sections Four and Three with all hands present or accounted for.

17 May – Moored starboard side to berth1 Port Louis, Mauritius with standard mooring lines doubled under the OPCON and ADCON of COMPACAREA. Ships status is Bravo. All deck, anchor, and aircraft warning lights are energized and burning brightly. Material Condition Yoke is set throughout the ship. Receiving potable water and sewage via barge and phone lines and electrical service via shore ties. All hands are on authorized liberty with the exception of Duty Section Four. Other ships present are commercial fishing vessels. 0630 Liberty expired for all hands. 0715 Pilot embarked Sherman. 0735 Set the Special Sea Detail. 0756 Clutched in both MDE's and placed in pilothouse control. 0803 Underway from Port Louis, Mauritius observing International navigation rules. 0817 Secured Special Sea Detail with the exception of the Navigation and Anchor DETIL. 0819 Placed both MDE's in engine room control. 0824 Secured Navigation and Anchor Detail. 0839 Set Flight Quarters Condition One for launch. 0847 Set the Helo Refueling Detail. 0849 Commenced cold refuel with CGNR-6596 on deck. 0857 Secured Helo Refueling Detail having transferred 52 gallons of JP-5. 0912 Helo off deck and away to port with three POB. 0915 Secured Flight Quarters Condition One, set Flight Con 2. 0928 Set Flight Quarters Condition One for landing. 0935 CGNR-6596 on deck with primary tie downs installed. 0938 Helo shut down engines and disengaged rotors. 0940 Secured from Flight Quarters Condition One. 1118 Traversed Helo into hanger with primary and secondary tie downs in place.

18 May - Underway in the Indian Ocean in position 017° 34.4S 054° 53.5E enroute Diego Suarez, Madagascar. Both MDE's are on line and turning for 15 knots. Material condition Yoke and Dog Zebra are set throughout the ship with EMCON DELTA in effect. SHERMAN is under the OPCON and TACON of COMPACAREA Alameda, CA. 0100 Retarded clocks to conform with -3 Charlie time zone, time now 0000 -3 Charlie.

19 May - Underway in the Indian Ocean in position 013° 03.2S 053° 28.7E enroute Diego Suarez, Madagascar. NR1 MDE is on line and turning for 12 knots. Material condition Yoke and Dog Zebra are set throughout the ship with EMCON DELTA in effect. SHERMAN is under the OPCON and TACON of COMPACAREA Alameda, CA. 0700 Set the Special Sea Detail. (part of log unreadable) 0825 Moored port side Antsiranana pier. 0826 Declutched both MDE's and placed in engine room control. 0841 Secured the Special Sea Detail 1200 Liberty granted for Duty Sections 2, 3 and 4 to expire 0845 20 May 2001.

20 May - Moored port side to commercial pier to Diego Suarez, Madagascar with standard mooring lines doubled under the OPCON and ADCON of COMPACAREA Alameda, CA. NR2 SSDG is providing electrical power. All deck, anchor and aircraft warning lights are energized and burning brightly. Material condition Yoke is set throughout the ship. Receiving telephone services via shore tie, Ships status is Bravo. All hands are on authorized liberty with the exception of duty section One. 0845 Liberty expired for duty section 1 and 2. Held morning muster all hands present or accounted for. <u>0907</u> Liberty granted to duty section 1 to expire 0530 21 May 2001.

21 May - Moored port side to commercial pier to Diego Suarez, Madagascar with standard mooring lines doubled under the OPCON and ADCON of COMPACAREA Alameda, CA. NR2 SSDG is providing electrical power. All deck, anchor and aircraft warning lights are energized and burning brightly. Material condition Yoke is set throughout the ship. Receiving telephone services via shore tie, Ships status is Bravo. All hands are on authorized liberty with the exception of duty section Two. <u>0625</u> Shifted watch from the quarterdeck to the pilothouse. <u>0632</u> Set the Special Sea Detail. <u>0705</u> Underway in Antsiranana Bay, Madagascar. <u>0718</u> Secured Special Sea Detail with the exception of Nav and Anchor Detail. <u>0744</u> Secured Anchor Detail. <u>0752</u> Secured Navigation Detail.

22 May – Underway in the Mozambique Channel in position 13° 04.4S 44° 21.7E. Both MDE's are on line turning for 15 knots. Material condition Yoke and Dog Zebra are set throughout the ship with EMCON DELTA in effect. SHERMAN is under the OPCON and TACON of COMPACAREA Alameda, CA. 0200 Retarded clocks back 1 hour to conform with -2 Bravo. <u>1440</u> Traversed Helo onto flight deck, primary and secondary tie downs in place. <u>1500</u> Set Flight Con 1, removed secondary tie downs. <u>1514</u> Helo started engines and engaged rotors. <u>1520</u> Commenced Hot Refuel on deck with Helo. <u>1523</u> Secured Hot Refuel on deck after having transferred 100 gallons of JP-5 to Helo. <u>1527</u> Helo off deck and away to port, 03 POB. <u>1550</u> Set Flight Con 4 for HIFR <u>1556</u> Helo over deck. 1558 Passed refueling hose to Helo. 1600 Helo away to port. 1604 Set Flight Con One for landing. 1609 Commenced touch and goes. 1613 Helo on deck with TALON engagement after having completed two touch and goes. 1624 Helo off deck and away to port 03 POB. 1625 Helo commenced touch and goes and landings. 1640 Helo on deck with TALON engagement after having completed 7 touch and go landings, primary tie downs in place. 1641 Helo off deck away to port. 1645 Helo on deck. 1650 Commenced touch and go landings. 1655 Secured after completing 03 touch and go landings. 1656 Commenced Hot Refuel on deck with Helo. 1658 Secured pump after transferring 76 gallons of JP-5 to Helo. 1700 Seured engines and rotors. 1703 Secured Flight Quarters. 1838 Traversed Helo into hanger, primary and secondary tie downs in place.

23 May – Underway in the Mozambique Channel in position 17° 20.6S 41° 45.8E. Both MDE's are on line turning for 15 knots. Material condition Yoke and Dog Zebra are set throughout the ship with EMCON DELTA in effect. SHERMAN is under the OPCON and TACON of COMPACAREA Alameda, CA. 1300 Commenced 1518 Commenced General Military Training on Alcohol/Drugs/Civil Rights/ Human Relations. 1430 Secured from GMT. MK75, 76mm Gunnery Exercise. 1542 Commenced Fire of 76mm. 1544 Secured from Gunnery Exercise having expanded 50 rounds of 76mm with no apparent casualties. 1548 Commenced CIWS PACFIRE. 1551 Commenced fire of CIWS. 1600 Secured from CIWS PACFIRE having expanded 300 rounds of 20mm with no apparent casualties.

24 May - Underway in the Mozambique Channel in position 22° 06.7S 38° 24.9E. Both MDE's are on line turning for 15 knots. Material condition Yoke and Dog Zebra are set throughout the ship with EMCON DELTA in effect. SHERMAN is under the OPCON and TACON of COMPACAREA Alameda, CA.

25 May - Underway in the Mozambique Channel in position 26° 52.3S 33° 53.3E. Enroute Cape town, South Africa. Both MDE's are on line turning for 15 knots. Material condition Yoke and Dog Zebra are set throughout the ship with EMCON DELTA in effect. SHERMAN is under the OPCON and TACON of COMPACAREA Alameda, CA. 0145 NR1 SSDG OOC. 0427 NR1 SSDG online. 0812 Traversed CGNR-6596 onto flight deck, primary and secondary tie downs in place. 0818 Set the Helo refueling Bill. 0834 Secured the Helo Refueling Bill having transferred 8? Gallons of JP-5 to CGNR-6596. 0840 Set Flight Quarters Condition 1 for launch. 0913 Secured from Flight Quarters Condition 1. 1000 Set Flight Quarters Condition 1 for launch. 1009 Removed secondary tie downs and engaged TALON hook on CGNR-6596. 1016 Helo started engines and engaged rotors. 1024 Removed primary tie downs and disengaged TALON hook from CGNR-6596. 1025 Helo off deck and away to port with 04 POB enroute Durban (Louis Botha) airport, South Africa. 1027 Secured Flight Quarters Condition 1, set Flight Quarters Condition 2. 1438 Set Flight Quarters Condition One for Landing. 1447 Helo is on deck with TALON engaged. 1450 Helo disengaged rotors and stopped engines. 1451 Secured from Flight Quarters Condition One. 1625 Traversed Helo into hanger with primary, secondary, and foul weather tie downs in place

26 May - Underway in the Mozambique Channel in position 31° 09.7S 30° 29.5E. Both MDE's are on line turning for 13.5 knots. Material condition Yoke and Dog Zebra are set throughout the ship with EMCON DELTA in effect. SHERMAN is under the OPCON and TACON of COMPACAREA Alameda, CA. 1120 Overheard vessel hailing port control for assistance in position 33° 24.3S 27° 42E, vessel identified as MODERN DRIVE. 1130 SHERMAN diverted to assist vessel MODERN DRIVE in position 33° 23.6S 27° 43.3E with engine room fire 20 people on board. 1141 M/V MODERN DRIVE request crew to be evacuated by Helo. 1148 M/V MODERN DRIVE reports fire under control crew no longer needs to be evacuated. 1149 M/V MODERN DRIVE reports 20 POB no engine room power, drifting, rolling up to 25°. 1153 M/V TRACER enroute to assist, M/V TRACER 6.1 NM from scene. 1225 M/V TRACER on scene with M/V MODERN DRIVE. 1338 SHERMAN on scene with M/V MODERN DRIVE and M/V TRACER in position 33° 22.8S 27° 43.1E. 1415 M/ MODERN DRIVE reports crew unable to access engine room to check status of fire due to smoke, heat, and smell of gasoline. 1445 M/V TRACER released to continue voyage by East London Port Control. 1445 Tug IMPUNZI underway from East London South, Africa enroute M/V MODERN DRIVE. 2000 Commercial tug IMPUNZI on scene with M/V MODERN DRIVE in position 33° 24.2S 27° 45.4E. 2134 East London Port Control recalled commercial tug IMPUNZI due to severe weather conditions. SHERMAN requests to remain on scene until another tug arrives.

27 May - Underway in the Indian Ocean in position 33° 23.6S 27° 48.1E. Both MDE's are on line turning for 5 knots. Material condition Yoke and Dog Zebra are set throughout the ship with EMCON DELTA in effect. SHERMAN is under the OPCON and TACON of COMPACAREA Alameda, CA. SHERMAN is currently on station with M/V MODERN DRIVE until commercial tug arrives. 0908 M/V MODERN DRIVE reported that it was unable to start generators. Secured emergency generators due to overheating. 0934 M/V MODERN DRIVE reported that emergency generator is running again and crew is about to inspect cargo hold. 0952 M/V MODERN DRIVE reported low lube oil pressure in its emergency generator. M/V MODERN DRIVE secured emergency generator. 1001 M/V MODERN DRIVE reported emergency generator is running intermittently due to low lube oil pressure. 1050 M/V MODERN DRIVE reported 4 holes on port side 5.5 meters above the water line in cargo hold NO.5 believed to be caused by shifting cargo. 1114 SHE-2 lowered to the main deck. 1125 SHE-2 lowered to the water and away to port with BM3 as coxswain SNBM, MK2 as boat crew and CWO Bellairs, SN, BM3 Alberici, and LTJG Strickland as passengers. 1135 SHE-2 alongside M/V MODERN DRIVE. 1200 All passengers safely aboard M/V MODERN DRIVE. 1212 Commenced initial inspection of M/V MODERN DRIVE. 1218 Set Towing Bill. 1311 Passed the messenger to M/V MODERN DRIVE via heaving line. 1424 Commenced towing M/V MODERN DRIVE. 1456 Helo over deck disembarking 03 POB on to M/V MODERN DRIVE. 1504 Helo away to port. 1510 Commenced preparations to break towing line with M/V MODERN DRIVE. 1515 Secured Towing Bill, SHE-2 alongside to port 1516 SHE-2 raised to main deck rail, boat crew safely aboard SHERMAN. 1656 SHE-2 lowered to water with BM3 as coxswain SNBM and MK2 as boat crew. 1658 SHE-2 away to port and alongside

M/V MODERN DRIVE. <u>1658</u> SHE-2 embarked 03 POB from M/V MODERN DRIVE. <u>1703</u> SHE-2 alongside SHERMAN 03 passengers safely aboard. <u>1709</u> SHE-2 cradled for sea boat crew safely aboard SHERMAN. <u>2242</u> Commenced preparations to break tow with M/V MODERN DRIVE. <u>2245</u> Commercial tug PENTOW SALVAR on scene in position 33° 21.3S 27° 51.0E. <u>2303</u> M/V MODERN DRIVE tested main propulsion, test satisfactory. <u>2327</u> M/V MODERN DRIVE commenced heaving on towing line. <u>2337</u> Tow broken with M/V MODERN DRIVE, secured towing lights and restricted inability to maneuver lights. <u>2354</u> Retrieved towing line from M/V MODERN DRIVE.

28 May - Underway in the Indian Ocean in position 33° 23.3S 27° 46.2E on scene with M/V MODERN DRIVE and tug Pentow Salvor. . Both MDE's are on line turning for 4 knots. Material condition Yoke and Dog Zebra are set throughout the ship with EMCON DELTA in effect. SHERMAN is under the OPCON and TACON of COMPACAREA Alameda, CA. <u>0024</u> Tug PENTOW SALVAR along to stern. <u>0026</u> Passed M/V MODERN DRIVE'S towing hawser to tug PENTOW SALVAR. <u>0030</u> Tug PENTOW SALVAR away. <u>0038</u> Secured the Towing Bill, shifted MDE's from pilot house to engine room control. <u>0047</u> USCGC SHERMAN underway from position 33° 31.1S 27° 41.1E. <u>1930</u> Commenced preparations for entering port in Table Bay, South Africa.

29 May - Underway in the Atlantic Ocean in position 34° 54.0S 19° 52.0E. NR1 MGT is on line turning for 22 knots. Material condition Yoke and Dog Zebra are set throughout the ship with EMCON DELTA in effect. SHERMAN is under the OPCON and TACON of COMPACAREA Alameda, CA. <u>0611</u> Traversed Helo onto flight deck with primary and secondary in downs in place. (Part of log unreadable) 0724 Helo off deck and away to port with 03 POB. 0727 Secured Flt Con 1, set Flt Con 2. 0755 Set Flight Con 1 for landing, 05 POB. 0800 CGNR-6596 on deck, disembarked 02 passengers. 0810 Commenced Hot Refuel on deck with Helo. 0820 Secured from refueling, 82 gallons of JP-5 transferred to Helo. 0821 Helo away to port, secured Flight Quarters. 0845 Set the Special Sea Detail. 0940 Pilot vessel alongside to port. 0941 Embarked Table Bay Harbor Pilot. 0956 Placed both engines in engine room control, lowered ad tested Bow Prop, test satisfactory, placed in pilot house control. 1015 Moored Jetty NO. 2 starboard side, Table Bay, South Africa. 1016 Declutched both MDE'S. 1017 Raised and secured Bow Prop. 1022 Secured the Special Sea Detail. 1027 OOD shifted the watch from the pilot house to the quarterdeck. 1215 Liberty granted to all hands with the exception of Duty Section Three. (Part of log unreadable).

30 May – Moored starboard side to NO.2 Jetty Cape Town South Africa with standard mooring lines doubled under the OPCON and TACON of COMPACAREA Alameda, CA. Ships status is Bravo-12. All deck, anchor, and aircraft warning lights are energized and burning brightly. NR? SSDG is providing electrical power. Receiving potable water, sewage, and telephone services via shore tie. Material condition Yoke is set throughout the ship. CGNR-6596 is on deck at Ysterplaat Air Base. All hands are on authorized liberty with the exception of Duty Section Three. 0815 Held muster for Duty Sections Three and Four. All hands present or accounted for. 0830 Liberty granted to Duty Section Three to expire onboard no later than 0830 02 June 2001.

31 May – Moored starboard side to NO.2 Jetty Cape Town South Africa with standard mooring lines doubled under the OPCON and TACON of COMPACAREA Alameda, CA. Ships status is Bravo-12. All deck, anchor, and aircraft warning lights are energized and burning brightly. NR? SSDG is providing electrical power. Receiving potable water, sewage, and telephone services via shore tie. Material condition Yoke is set throughout the ship. CGNR-6596 is on deck at Ysterplaat Air Base. All hands are on authorized liberty with the exception of Duty Section Four. 0815 Held muster for Duty Sections One and Four. All hands present or accounted for. 0830 Liberty granted to Duty Section Four to expire onboard no later than 0830 02 June 2001.

01 June – Moored starboard side to NO.2 Jetty Cape Town South Africa with standard mooring lines doubled under the OPCON and TACON of COMPACAREA Alameda, CA. Ships status is Bravo-12. All deck, anchor, and aircraft warning lights are energized and burning brightly. NR? SSDG is providing electrical power. Receiving potable water, sewage, and telephone services via shore tie. Material condition Yoke is set throughout the ship. CGNR-6596 is on deck at Ysterplaat Air Base. All hands are on authorized liberty with the exception of Duty Section One. 0815 Held muster for Duty Sections One and Two. All hands present or accounted for. 0830 Liberty granted to Duty Section One to expire onboard no later than 0830 02 June 2001.

02 June – Moored starboard side to NO.2 Jetty Cape Town South Africa with standard mooring lines doubled under the OPCON and TACON of COMPACAREA Alameda, CA. Ships status is Bravo-12. All deck, anchor, and aircraft warning lights are energized and burning brightly. NR? SSDG is providing electrical power. Receiving potable water, sewage, and telephone services via shore tie. Material condition Yoke is set throughout the ship. CGNR-6596 is on deck. All hands are on authorized liberty with the exception of Duty Section Two. <u>0830</u> Liberty expired for all hands. <u>0842</u> Singled up all mooring lines. <u>0930</u> Set the Special Sea Detail. <u>0937</u> Lowered and tested Bow Prop, test satisfactory placed in pilot house control. <u>1000</u> Set the Restricted Maneuvering Bill. <u>1014</u> Clutched in both MDE's placed in pilot house control. <u>1022</u> Underway in Table Bay, South Africa. <u>1048</u> Moored Duncan Dock Berth E Cape Town, South Africa. <u>1059</u> Secured the Special Sea Detail. <u>1159</u> Commenced Fueling draft FWD 12'0" AFT 15'6". <u>1353</u> Secured from fueling having received 78,752 gallons of JP-5 fuel, drafts 15'0" FWD and 15' 0" AFT. <u>1441</u> Set the Special Sea Detail. <u>1515</u> Underway in Table Bay, South Africa. <u>1530</u> Raised and secured Bow Prop. <u>1532</u> Secured Special Sea Detail with exception of Nav and Anchor Detail. Disembarked Table Bay Pilot. <u>1533</u> Pilot Boat away to starboard. <u>1536</u> Secured Anchor Detail, placed both MDE's in engine room control. <u>1617</u> Secured Navigation Detail. <u>1620</u> Removed primary and secondary tie downs from CGNR-6596 <u>1621</u> Traversed Helo to hanger primary and secondary tie downs in place.

03 June - Underway in the Atlantic Ocean in position 32° 28.4S 16° 51.3E in route Cape Verde. Both MDE's are on line turning for 15 knots. Material condition Yoke and Dog Zebra are set throughout the ship with EMCON DELTA in effect. SHERMAN is under the OPCON and TACON of COMPACAREA Alameda, CA.

04 June - Underway in the Atlantic Ocean in position 25° 27.4S 12° 47.0E in route Cape Verde. Both MDE's are on line turning for 15 knots. Material condition Yoke and Dog Zebra are set throughout the ship with EMCON DELTA in effect. SHERMAN is under the OPCON and TACON of COMPACAREA Alameda, CA. 0200 Retarded clocks 1 hour to conform to -1 Alpha, current time is 0100. 1 enroute 1320 Boat crew embarked SHE-2, boat crew consists of BM3, BN3 and MK3. 1324 SHE-2 lowered to the water and away to port enroute Man Overboard Drill. 1335 SHE-2 alongside to port, boat crew safely onboard SHERMAN. 1338 SHE-2 secured to port 01 deck rail. 1354 SHE-2 lowered to the water. 1355 SHE-2 away to port. 1357 SHE-2 alongside to port. 1358 SHE-2 raised to the rail boat crew safely on board. 1407 SHE-2 cradled and secured for sea.

05 June - Underway in the Atlantic Ocean in position 21° 51.6S 10° 41.7E in route Cape Verde. Both MDE's are on line turning for 15 knots. Material condition Yoke and Dog Zebra are set throughout the ship with EMCON DELTA in effect. SHERMAN is under the OPCON and TACON of COMPACAREA Alameda, CA.

06 June - Underway in the Atlantic Ocean in position 16° 18.0S 08° 29.5E in route Cape Verde, Africa. Both MDE's are on line turning for 15 knots. Material condition Yoke and Dog Zebra are set throughout the ship with EMCON DELTA in effect. SHERMAN is under the OPCON and TACON of COMPACAREA Alameda, CA. 1115 Traversed CGNR-6596 onto deck with primary and secondary tie downs in place. 1300 Commenced LE Training. 1400 Secured from LE training. 1500 Commenced Commanding Officers Non-Judicial Proceedings under article 15 UCMJ. <u>1523</u> Secured from Commanding Officers Non-Judicial Proceedings with the following results. MK2, USCG is awarded reduction to E-4 for violation of article 92 UCMJ. QM3, USCG is awarded reduction to E-3 for violation of article 92 UCMJ. <u>1603</u> Commenced Man Overboard Drill <u>1607</u> SHE-2 lowered to the water with BM2 as coxswain, BM3 and SN as boat crew. <u>1608</u> (part of log unreadable). <u>1643</u> Secured from Man Overboard Drill.

07 June - Underway in the Atlantic Ocean in position 11° 01.6S 05° 40.8E in route Cape Verde, Africa. Both MDE's are on line turning for 15 knots. Material condition Yoke and Dog Zebra are set throughout the ship with EMCON DELTA in effect. SHERMAN is under the OPCON and TACON of COMPACAREA Alameda, CA. 0200 Retarded clocks back 1 hour to conform with Greenwich Mean Time. <u>0534</u> Commenced General Emergency Drill. <u>0638</u> Secured from General Emergency Drill. <u>1551</u> Traversed CGNR-6596 onto the flight deck, primary and secondary tie downs in place. <u>1630</u> Set Flight Quarters Condition 1 for launch, removed secondary tie downs from Helo. <u>1634</u> Set the Helo Refueling Detail. <u>1638</u> Secured the Helo Refueling Bill having transferred 45 gallons of JP-5 to Helo. <u>1642</u> Helo started engines and engaged rotors. <u>1657</u> Removed primary tie downs from Helo. <u>1658</u> Helo is off the deck and away to port with 03 POB. <u>1704</u> Secured Flight Con 1, set Flight Com 2. <u>1748</u> Set Flight Con 3 for VERTREP with Helo. <u>1754</u> Secured VERTREP, Helo away to port. <u>1755</u> Helo over the deck, commenced VERTREP. <u>1756</u> Secured VERTREP, Helo away to port, secured Flight Con 3, set Flight Con 1. <u>1801</u> Helo is on deck, TALON engaged. 1804 Commenced Hot Refuel on deck with Helo <u>1806</u> Secured Hot

Refuel having transferred 71 gallons of JP-5 to Helo. <u>1809</u> SHE-2 lowered to the water with BM1 as coxswain, BM2, MK3, SNMB, SN, AND SN as boat crew. <u>1812</u> SHE-2 away to port. <u>1827</u> SHE-2 alongside to port. <u>1828</u> SHE-2 raised to the main deck rail, coxswain and boat crew safely aboard SHERMAN. <u>1837</u> Disengaged TALON. <u>1838</u> Helo off the deck and away to port with 06 POB. <u>1840</u> SHE-2 cradled, determined to be OOC, SHE-1 is the ready boat. <u>1842</u> ETR of SHE-2 unknown. <u>1853</u> SHE-2 repaired, SHE-2 is the ready boat. <u>1858</u> Commenced touch and go landings. <u>1901</u> Helo is on deck having completed 01 touch and go landing, TALON engaged. <u>1906</u> Helo secured engines and disengaged rotors, secured Flight Con 1, set the Helo Refueling Bill. <u>1909</u> Secured the Helo Refueling Bill having transferred 69 gallons of JP-5 to Helo. <u>1920</u> Installed primary tie downs on the Helo. <u>2051</u> Traversed Helo into hanger with primary and secondary tie downs in place.

08 June - Underway in the Atlantic Ocean in position 06° 57.0S 03° 31.5E in route Cape Verde, Africa. Both MDE's are on line turning for 15 knots. Material condition Yoke and Dog Zebra are set throughout the ship with EMCON DELTA in effect. SHERMAN is under the OPCON and TACON of COMPACAREA Alameda, CA. <u>1530</u> Commenced Commanding Officers Non-Judicial Proceedings. <u>1551</u> Secured from Commanding Officers Non-Judicial Proceedings with the following results. SA is awarded reduction to E-1 for violation of Article 121 of the UCMJ.

09 June - Underway in the Atlantic Ocean in position 06° 57.0S 03° 31.5E in route Cape Verde, Africa. Both MDE's are on line turning for 15 knots. Material condition Yoke and Dog Zebra are set throughout the ship with EMCON DELTA in effect. SHERMAN is under the OPCON and TACON of COMPACAREA Alameda, CA. 0701 Crossed the Equator at Prime Meridian 0115 SHE-2 lowered to (part of log unreadable) FN injured during swim call resulting in a lacerated right eyelid, SNM is fir for full duty. 0532 Commenced General Quarters Drill. 0610 Secured from General Quarters Drill. 0807 SHE-2 alongside to port. 0808 SHE-2 raised to the main deck rail, both MDE's clutched in. 0809 Coxswain and boat crew safely aboard. 0811 SHE-2 cradled for sea.

Note – Emerald Shellback

10 June - Underway in the Atlantic Ocean in position 01° 16.7N 04° 22.4W in route Cape Verde, Africa. Both MDE's are on line turning for 16 knots. Material condition Yoke and Dog Zebra are set throughout the ship with EMCON DELTA in effect. SHERMAN is under the OPCON and TACON of COMPACAREA Alameda, CA.

11 June - Underway in the Atlantic Ocean in position 03° 09.4N 10° 46.7W in route Cape Verde, Africa. Both MDE's are on line turning for 16 knots. Material condition Yoke and Dog Zebra are set throughout the ship with EMCON DELTA in effect. SHERMAN is under the OPCON and TACON of COMPACAREA Alameda, CA. 0200 Retarded clocks 1 hour to conform with +1 November. New time 0100. 1530 Commenced Gunnery Exercise for 76 and CWIS PACFIRE. 1631 Secured from Gunnery Exercise with no apparent casualties after expanding 31 rounds for 76mm and 300 rounds for CWIS.

12 June - Underway in the Atlantic Ocean in position 06° 04.9N 16° 02.8W in route Cape Verde, Africa. Both MDE's are on line turning for 16 knots. Material condition Yoke and Dog Zebra are set throughout the ship with EMCON DELTA in effect. SHERMAN is under the OPCON and TACON of COMPACAREA Alameda, CA. 0755 Test results for JP-5 SED-1, FW-5, and FSH .17. 1330 Helo traversed to flight deck. 1340 Set Flight Con 1 for launch. 1350 Helo started engines and engaged rotors. 1406 CGNR-6596 above deck and clear to port. 1407 Commenced Helo Crash at Sea Drill. 1409 SHE-2 lowered to the water with BM2 as coxswain, SN, MK3 and BM3 as boat crew. 1416 SHE-2 raised to main deck rail, secured from Helo Crash at Sea Drill. 1418 SHE-2 cradled for sea. 1445 Set Flight Con 1 for landing. 1456 Helo on deck, primary tie downs in place. 1457 Commenced Hot Refuel on deck with Helo. 1458 Engaged pump. 1500 Secured Hot Refueling after having transferred 73 gallons of JP-5 to Helo. (part of log unreadable) 1529 Helo on deck. 1539 Helo off deck and away to port with 04 POB. 1542 CGNR-6596 commenced touch and go landings. 1602 CGNR-6596 off deck and away to port having completed 12 touch and go landings. 1604 CGNR-6596 on deck, primary tie downs in place. 1605 Helo secured engines and disengaged rotors. 1607 Commenced Helo Crash on Deck Drill. 1622 Set the Helo Refueling Bill, secured Flight Quarters Condition 1. Secured Helo Refueling Bill having transferred 80 gallons of JP-5 to Helo. 1638 Secured Helo Crash on Deck Drill. 1841 Traversed Helo to hanger, primary and secondary tie downs in place.

13 June - Underway in the Atlantic Ocean in position 10° 43.5N 20° 19.8W in route Cape Verde, Africa. Both MDE's are on line turning for 16 knots. Material condition Yoke and Dog Zebra are set throughout the ship with EMCON DELTA in effect. SHERMAN is under the OPCON and TACON of COMPACAREA Alameda, CA. <u>1432</u> Commenced preparations for entering port in Mindelo, Cape Verde.

14 June - Underway in the Atlantic Ocean in position 15° 29.1N 24° 27.5W in route Cape Verde, Africa. Both MDE's are on line turning for 15 knots. Material condition Yoke and Dog Zebra are set throughout the ship with EMCON DELTA in effect. SHERMAN is under the OPCON and TACON of COMPACAREA Alameda, CA. <u>0741</u> Pilot vessel alongside to starboard. <u>0743</u> Embarked Sau Vincente pilot. <u>0755</u> Moored port side Berth 2 Pontinha Pier. <u>0800</u> Secured Special Sea Detail, with the exception of the Brow Detail. <u>0900</u> Commence fueling drafts FWD 13' 4" AFT 15. <u>1013</u> Liberty is granted to all hands with the exception of Duty Section Three to expire 0845 15 June 2001 for Duty Section Four and 0830 16 June 2001 for all hands. <u>1309</u> Secured from fueling, total gallons 76,259, drafts FWD 14' 6" AFT 14' 6".

15 June - Moored portside to Berth 2 Pontinha Pier, Mindelo, Cape Verde with all standard mooring lines doubled under the OPCON and ADCON of COMPACAREA, Alameda, CA. Ships status is Bravo-6. All deck, anchor, and aircraft warning lights are energized and burning brightly. Material condition Yoke is set throughout the ship. NR2 SSDG is providing electrical power. Receiving potable water via shore tie. All hands are on authorized liberty with the exception of Duty Section Three. <u>0845</u> Held muster for Duty Sections Three and Four, all hands present or accounted for. <u>0923</u> Liberty granted to Duty Section Three to expire0830 16 June 2001.

16 June - Moored portside to Berth 2 Pontinha Pier, Mindelo, Cape Verde with all standard mooring lines doubled under the OPCON and ADCON of COMPACAREA, Alameda, CA. Ships status is Bravo-6. All deck, anchor, and aircraft warning lights are energized and burning brightly. Material condition Yoke is set throughout the ship. NR2 SSDG is providing electrical power. Receiving potable water via shore tie. All hands are on authorized liberty with the exception of Duty Section Four. 0830 Liberty expired for all hands. 0912 Singled up on all mooring lines. 0930 Set the Special Sea Detail. 0944 Clutched in both MDE's, placed in pilothouse control. 0945 Lowered and tested Bow Prop, test satisfactory, placed in pilothouse control. 0952 Underway in the North Atlantic Ocean. 1001 Raised and secured Bow Prop. 1003 Placed both MDE'S in engine room control. 1009 Secured the Special Sea Detail with the exception of the Navigation and Anchor Detail. 1012 Secured the Navigation and Anchor Detail. 1820 Experienced casualty with X-Band radar, estimated time to repair unknown.

17 June - Underway in the North Atlantic Ocean in position 16° 12.5N 28° 10.6W in route Bridgetown, Barbados. Both MDE's are on line turning for 15 knots. Material condition Yoke and Dog Zebra are set throughout the ship with EMCON DELTA in effect. SHERMAN is under the OPCON, ADCON and TACON of COMPACAREA Alameda, CA. 0200 Retarded clocks one hour to conform to +2 Oscar Time now 0100. 1205 Commenced Fish Call. 1600 Secured from Fish Call.

18 June - Underway in the North Atlantic Ocean in position 15° 00.1N 33° 54.2W in route Bridgetown, Barbados. Both MDE's are on line turning for 15 knots. Material condition Yoke and Dog Zebra are set throughout the ship with EMCON

DELTA in effect. SHERMAN is under the OPCON, ADCON and TACON of COMPACAREA Alameda, CA. 0906 Commenced BECCE Drills. 0933 Secured from BECCE Drills. 0942 Traversed Helo onto flight deck, primary and secondary tie downs in place. 1305 Commenced BECCE Drills. 1338 Secured from BECCE Drills. 1540 Commenced Man Overboard Drill. 1618 Secured from Man Overboard Drill. 1635 Traversed Helo into hanger, primary, secondary, and heavy weather tie downs in place.

19 June - Underway in the North Atlantic Ocean in position 14° 33.8N 39° 48.9W in route Bridgetown, Barbados. Both MDE's are on line turning for 15 knots. Material condition Yoke and Dog Zebra are set throughout the ship with EMCON DELTA in effect. SHERMAN is under the OPCON, ADCON and TACON of COMPACAREA Alameda, CA. 0200 Retarded ships clocks to conform with +3 Papa Time Zone. Time now 0100 +3 Papa. 1541 Commenced Man Overboard Drill. 1550 Secured from Man Overboard Drill.

20 June - Underway in the North Atlantic Ocean in position 14° 13.2N 46° 49.1W in route Bridgetown, Barbados. Both MDE's are on line turning for 15 knots. Material condition Yoke and Dog Zebra are set throughout the ship with EMCON DELTA in effect. SHERMAN is under the OPCON, ADCON and TACON of COMPACAREA Alameda, CA. 0905 Commenced BECCE Drills. 1000 Secured from BECCE Drills. 1305 Commenced BECCE Drills. 1342 Secured from BECCE Drills, commenced General Emergency Drill. 1433 Secured from General Emergency Drill.

21 June - Underway in the North Atlantic Ocean in position (position left blank on log) in route Bridgetown, Barbados. Both MDE's are on line turning for 15 knots. Material condition Yoke and Dog Zebra are set throughout the ship with EMCON DELTA in effect. SHERMAN is under the OPCON, ADCON and TACON of COMPACAREA Alameda, CA. 0017 Shifted to MK27 Gyro Compass. 0200 Retarded clocks 1 hour to conform to +4 Quebec. Time on deck 0100 +4 Quebec. 1030 Commenced Steering Casualty Drill. 1046 Secured from Steering Casualty Drill. 1126 Commenced preparations for entering port in Bridgetown, Barbados.

22 June - Underway in the North Atlantic Ocean in position 13° 10.2N 57° 55.7W in route Bridgetown, Barbados. Both MDE's are on line turning for 15 knots. Material condition Yoke and Dog Zebra are set throughout the ship with EMCON DELTA in effect. SHERMAN is under the OPCON, ADCON and TACON of COMPACAREA Alameda, CA. 0701 Set the Navigation and Anchor Details. 0723 Set the Special Sea Detail. 0748 Pilot Vessel alongside to port, pilot safely aboard. 0749 Pilot Vessel away to port. 0808 Moored starboard side to Bridgetown, Barbados. 0809 Declutched both MDE's. 0849 Shifted the watch from the pilothouse to the quarterdeck. 1000 Liberty is granted to all hands with the exception of Duty Section One.

23 June – Moored starboard side to Bridgetown, Barbados with standard mooring lines doubled under the OPCON, and ADCON of COMPACAREA Alameda, CA. Ships status is Bravo-12. All deck, anchor, and aircraft warning lights are energized and burning brightly. Material condition Yoke is set throughout the ship. Receiving potable water, sewage via shore tie. All hands are on authorized liberty with the exception of Duty Section One. <u>0845</u> Liberty expired for duty section Two. Held morning muster for Duty Section One and Two, all present or accounted for. Liberty granted to Duty Section One to expire 0830 26 June 2001. <u>1440</u> TC3 arrived PCS from TRACEN Petaluma. SA arrived PCS from TRACEN Cape May. FN arrived PCS from TRACEN Petaluma. SNFS arrived PCS from TRACEN Petaluma. SA arrive PCS from TRACEN Cape May.

24 June – Moored starboard side to Bridgetown, Barbados with standard mooring lines doubled under the OPCON, and ADCON of COMPACAREA Alameda, CA. Ships status is Bravo-12. All deck, anchor, and aircraft warning lights are energized and burning brightly. Material condition Yoke is set throughout the ship. Receiving potable water, sewage via shore tie. All hands are on authorized liberty with the exception of Duty Section Two. <u>0845</u> Liberty expired for duty section Three. Held morning muster for Duty Section Two and Three, all present or accounted for. Liberty granted to Duty Section Two to expire 0830 26 June 2001. 1557 SA reported PCS from TRACEN Petaluma. 2037 ENS, ENS and ENS reported PCS from Coast Guard Academy. SA reported PCS from TRACEN Cape May.

25 June – Moored starboard side to Bridgetown, Barbados with standard mooring lines doubled under the OPCON, and ADCON of COMPACAREA Alameda, CA. Ships status is Bravo-12. All deck, anchor, and aircraft warning lights are energized and burning brightly. Material condition Yoke is set throughout the ship. Receiving potable water, sewage via shore tie. All hands are on authorized liberty with the exception of Duty Section Three. <u>0845</u> Liberty expired for duty section Four.

Held morning muster for Duty Section Three and Four, all present or accounted for. Liberty granted to Duty Section Three to expire 0830 26 June 2001.

26 June – Moored starboard side to Bridgetown, Barbados with standard mooring lines doubled under the OPCON, and ADCON of COMPACAREA Alameda, CA. Ships status is Bravo-12. All deck, anchor, and aircraft warning lights are energized and burning brightly. Material condition Yoke is set throughout the ship. Receiving potable water, sewage via shore tie. All hands are on authorized liberty with the exception of Duty Section Four. 0830 Liberty expired for all hands. ???? Set the Special Sea Detail. 1002 Pilot safely aboard SHERMAN. 1004 Lowered and tested Bow Prop, test satisfactory. 1006 Clutched in both MDE's and placed in pilothouse. 1028 Underway from Bridgetown, Barbados. GM2 missed movement. 1043 Pilot vessel, away to starboard. Late Entry – 1040 Raised and secured Bow Prop. 1047 Placed both MDE's in engine room control. Late Entry – 0918 Traversed Helo onto flight deck, primary and secondary tie downs in place. 1439 Set Flight Quarters Condition 1 for launch. 1443 Removed primary and secondary tie downs from Helo. 1452 CGNR-6596 started engine and engaged rotors. 1505 Helo above deck and away to port. 1514 CGNR-6596 on deck, TALON engaged. 1515 Commenced Hot Refuel on deck with Helo, energized pump. 1518 Secured pump after transferring 61 gallons of JP-5 to Helo. ???? Helo off deck, secured Flight Quarters Condition 1, set Flight Quarters Condition 2. 1650 Set Flight Quarters Condition 1 for landing. 1702 CGNR-6596 on deck, TALON engaged primary tie downs in place. 1704 Commenced Hot Refuel on deck with Helo, engaged pump. 1708 Pump secured after transferring 128 gallons of JP-5 to Helo. 1710 Helo secured engines and disengaged rotors. 1712 Secured from Flight Quarters Condition 1. 1718 installed secondary tie downs on Helo. 1805 Entered Territorial Waters of St. Lucia. 1819 Entered Territorial Water of St. Vincent. 1834 Commenced Preparations for entering port in Oranjestad, Aruba. 1843 Exited Territorial Waters of St. Lucia. 185 Exited Territorial Waters of St. Vincent

27 June - Underway in the North Atlantic Ocean in position 13° 27.1N 62° 31.5W in route Aruba. Both MDE's are on line turning for 16 knots. Material condition Yoke and Dog Zebra are set throughout the ship with EMCON DELTA in effect. SHERMAN is under the OPCON, ADCON and TACON of COMPACAREA Alameda, CA. 0740 Removed secondary and primary tie downs from Helo. ???? Helo off deck and clear to port with 04 POB. 0828 Secured Flight Con 1, set Flight Con 2. 0830 Commenced preparations for entering port in Oranjestad, Aruba. 0942 Set Flight Con 1 for landing. 1000 CGNR-6596 on deck, TALON engaged, primary and secondary tie downs in place. Disembarked 01 POB. 1001 Commenced Hot Refuel on deck with Helo, pump energized. 1006 Secured Hot Refuel after transferring 158 gallons of JP-5 to Helo. 1007 Helo secured engines and disengaged rotors. 1010 Secured Flight Con 1. 1750 Set Flight Quarters Condition One for launch, removed secondary tie downs. 1808 Helo started engines and engaged rotors. 1810 Removed primary tie downs from Helo. 1811 Helo is off Deck and away to port. 1813 Secured from Flight Quarters Condition One, set Flight Quarters Condition Two. 1930 Set Flight Quarters Condition 1 for landing. 1942 Helo is on deck, primary tie downs in place. 1945 Helo secured engines and disengaged rotors. 1946 Secured Flight Con 1, set the Helo Refueling Bill. 1952 Secured the Helo Refueling Bill having transferred (no amount given) gallons of JP-5 to Helo. 2133 Traversed CGNR-6596 into hanger, primary, secondary, and heavy weather tie downs in place.

28 June - Underway in the Caribbean Sea in position 12° 48.3N 68° 53.2W in route Aruba. NR2 MDE is on line turning for 10 knots. Material condition Yoke and Dog Zebra are set throughout the ship with EMCON DELTA in effect. SHERMAN is under the OPCON, ADCON and TACON of COMPACAREA Alameda, CA. 0715 Set the Navigation and Anchor Detail. 0735 Placed both MDE's in pilothouse control. 0742 Pilot Vessel alongside to port. 0745 Embarked Oranjestad Pilot. 0751 Lowered and tested Bow Prop, test satisfactory, placed in pilothouse control. 0802 Moored portside Berth 2 Oranjestad, Aruba. 0807 Declutched both MDE's, placed in engine room control. 0811 Raised and secured Bow Prop. 0844 Shifted watch from the pilothouse to the quarterdeck. 0944 Liberty granted to all hands with the exception of Duty Section One to expire 0845 29 June 2001 for Duty Section Two and 1030 30 June 2001 for all hands. 1200 Sounded eight bells. 1236 Secured the Fueling Detail having taken on 28, 238 gallons of diesel fuel.

29 June – Moored portside to Berth 2 Oranjesad, Aruba with standard mooring lines doubled under the OPCON, ADCON and TACON of COMPACAREA Alameda, CA. Ships status is Bravo-12. All deck, anchor and aircraft warning lights are energized and burning brightly. NR2 SSDG is providing electrical power. All hands are on authorized liberty with the exception of Duty Section One. 0845 Liberty expired for Duty

Section Two. Held muster for Duty Sections One and Two. All hands present or accounted for. <u>0900</u> Liberty Granted to Duty Section Two to expire 1030 30 June 2001.

30 June – Moored portside to Berth 2 Oranjesad, Aruba with standard mooring lines doubled under the OPCON, ADCON and TACON of COMPACAREA Alameda, CA. Ships status is Bravo-12. All deck, anchor and aircraft warning lights are energized and burning brightly. NR2 SSDG is providing electrical power. All hands are on authorized liberty with the exception of Duty Section One. <u>1030</u> Liberty expired for all hands, Held Quarters for all hands, all hands present or accounted for. Singled up on all mooring lines. <u>1135</u> Set the Special Sea Detail. <u>1150</u> Bow Prop lowered and tested, test satisfactory, placed in pilothouse control. <u>1153</u> Clutched in both MDE's, placed in pilothouse control. <u>1158</u> Underway Oranjestad Harbor, Aruba. <u>1205</u> Raised and secured Bow Prop, embarked Oranjestad harbor Pilot. <u>1214</u> Placed both MDE's in engine room control. <u>1216</u> Secured Special Sea Detail. Pilot Vessel alongside to starboard, disembarked Pilot. <u>1428</u> Removed foul weather and secondary tie downs from Helo. <u>1431</u> CGNR-6596 Traversed to flight deck. <u>1452</u> Set Flight Con 1. <u>1509</u> Helo started engines and engaged rotors. 1527 Helo off deck and away to port, 04 POB. <u>1556</u> maneuvered to intercept M/V MANILA PRIDE 6 in position 12° 47.8N 70° 42.3W. 1557 Commenced hailing M/V MANILA PRIDE 6 on VHF channel 16 <u>1611</u> On-scene with M/V MANILA PRIDE 6 in position 12° 46.9N 70° 39.6W on course 090° true 15 knots. <u>1618</u> M/V MANILA PRIDE 6 responded to hailing on VHF 16. Commenced Query. <u>1643</u> Set Flight Con 1 for landing. (Part of log unreadable) <u>1700</u> Energized pump. <u>1704</u> Secured pump after transferring 146 gallons of JP-5 to Helo. <u>1711</u> Helo commenced engine and rotor shutdown. <u>1854</u> Traversed Helo into hanger primary, secondary, and heave weather tie downs in place.

01 July - Underway in the Caribbean Sea in position 12° 51.0N 72° 21.8W using International Navigation Rules in transit to San Diego. Ship's course is 246° true, 255° magnetic. Both MDE'S are online turning for 16.5 knots. Both MGT's are in standby, NR2 SSDG is providing electrical power. All standard navigation lights are energized and burning brightly. Dog Zebra and material condition Yoke set throughout the ship with EMCON Delta in effect. Both boats are cradled for sea with SHE-2 as the ready boat. CGNR-6596 is hangered with primary, secondary, and heavy weather tie-downs in place. Sherman is under the OPCON, ADCON, and TACON of COMPACAREA Alameda. <u>0000</u> Late Entry, GM2 absent without leave <u>0048</u> Secured navigation lights by order of the commanding officer. <u>0200</u> Retarded clocks 1 hour to conform with +5 Romeo time zone, new time 0100 + Romeo. <u>0225</u> Energize all standard navigational lighting to comply with International rules of the road. <u>0248</u> Secured navigation lights by order of the commanding officer. <u>0520</u> Placed both SSDG'S in parallel. <u>0533</u> Observed sunrise. <u>0806</u> Traversed CGNR-6596 to flight deck <u>0836</u> Set Flight Con 1 <u>0847</u> Helo started engines and engaged rotors <u>0905</u> Removed primary tie downs from CGNR-6596, Helo above deck and away to port <u>0907</u> Secured Flight Con 1, set Flight Con 2. <u>0912</u> Clutched in both MGT's, declutched both MDE's, placed in immediate standby <u>1022</u> Declutched both MGT's, clutched in both MDE's <u>1030</u> Set Flight Con 1 <u>1045</u> Helo on deck, primary tie downs in place

1046 Commenced Hot Refuel on deck with Helo 1048 Energized pump 1051 Secured from Hot Refuel after transferring 141 gallons of JP-5 to Helo 1055 Helo commenced engine an rotor shut down 1056 Secured Flight Con 1 1103 All small arms and small arms ammunition present and accounted for 1200 Conducted test of ships alarms and air whistle, test satisfactory 1610 Set Flight Con 1 for launch 1625 CGNR-6596 started engines and engaged rotors, primary tie downs removed 1633 CGNR-6596 above deck and away to port 1636 Secured Flight Con 1, set Flight Con 2 1751 Set Flight Quarters Con 1 for landing 1801 CGNR-6596 on deck, primary tie downs in place 1804 Secured Flight Quarters Condition On, set the Helo Refueling Bill 1810 Secured the Helo Refueling Bill after having transferred 125 gallons of JP-5 to the Helo 1837 Observed sunset 1953 Clutched in NR2 MGT 1955 Declutched both MDE's, placed in standby 2028 Singled up on NR 1 SSDG 2041 Made preparations for entering the Panama Canal

02 July - Underway in the Caribbean Sea in position 10° 11.8N 78° 41.7W using International Navigation Rules, enroute San Diego via the Panama Canal. Ship's course is 250° true, 253° magnetic. NR 2 MGT is online turning for 22 knots. Both MDE's and NR1 MGT are in standby, NR1 SSDG is providing electrical power. All standard navigation lights are secured by order of the commanding officer. Dog Zebra and material condition Yoke set throughout the ship with EMCON Delta in effect. Both boats are cradled for sea with SHE-2 as the ready boat. CGNR-6596 is hangered with primary, secondary, and heavy weather tie-downs in place. Sherman is under the OPCON, ADCON, and TACON of COMPACAREA Alameda. 0000 Late Entry, GM2 absent without leave 0105 Clutched in both MDE's, declutched NR2 MGT 0230 Energized all standard navigation lights to conform with the rules of the road 0530 Set the Navigation and Anchor Detail 0555 Commenced approach on NW Outer Anchorage, Panama Canal Entrance 0603 Observed sunrise, secured all standard navigation lights 0618 Placed both MDE's in Pilot House control 0635 Anchored in Anchorage D, Panama Canal Entrance in position 09° 22.3N 79° 54.5W with 1 shot of chain on deck to the port anchor to a rocky bottom in 21.1 feet of water using the following bearings 316° True to Bravo, 157° to Foxtrot, 198° to Gulf and 2345 to Gulf, raised dayshapes, secured Navigation and Anchor Detail 0813 All small arms and ammunition present or accounted for 0934 M/V Albacore II alongside to starboard, 03 boarding officers safely aboard 0936 M/V Albacore II away to starboard 1003 M/V Jebel alongside to starboard, 03 boarding officers safely on board M/V Jebel 1005 M/V Jebel away to starboard 1045 SHE-2 lowered to the main deck rail, coxswain and boat crew safely aboard 1047 SHE-2 lowered to the water and away to port with BM3 as coxswain SN, MK3 and CWO as boat crew 1102 SHE-2 alongside to port 1105 SHE-2 raised to the main deck rail, coxswain and boat crew safely on board 1115 SHE-2 cradled for sea 1641 Set the Navigation and Anchor Detail 1642 M/V Pike alongside to starboard with pilot 1645 Pilot safely aboard, M/V Pike away to starboard 1702 Commenced heaving around on port anchor, clutched in both MDE's, placed in Pilot House control 1706 Anchors aweigh in Anchorage Area D, Panama Canal Entrance. 1711 Placed both MDE's in Engine Room control, howsed port anchor 1745 Placed both MDE's in Pilot House control 1751 Commenced approach on Gatun Lochs 1759 Cables attached port and starboard sides 1825 Closed locks with M/V Artic Ice inside locks. Commenced flooding first lock <u>1838</u> Observed sunset, energized all standard navigation lights and steering light <u>1841</u> Secured flooding of first lock, opened forward doors <u>1850</u> Closed rear doors, commenced flooding lock <u>1901</u> Opened forward doors to lock, maneuvered inside third lock <u>1908</u> Closed rear gates to lock, commenced flooding <u>1920</u> Commenced opening forward doors <u>1924</u> Removed port and starboard cables <u>1925</u> Departed from Gatun Lochs <u>1928</u> Raised and secured Bow Prop <u>1930</u> Placed both MDE's in Engine Room control <u>2138</u> Placed both MDE's in Pilot House control <u>2140</u> Lowered and tested Bow Prop, test satisfactory, placed in Pilot House Control <u>2142</u> Vessel alongside to starboard, line handlers safely aboard Sherman, vessel away to starboard <u>2216</u> Entered Pedro Miguel Lochs <u>2223</u> Passed and attached fore and aft cables <u>2241</u> Closed rear doors, began draining lock <u>2252</u> Opened front gates <u>2258</u> Raised and secured Bow Prop <u>2305</u> Lowered and tested Bow Prop <u>2320</u> Entered Miraflores Lochs <u>2323</u> Received and attached fore and aft cables <u>2343</u> Closed rear doors <u>2347</u> Opened front doors, began draining lock

3 July - Underway in the Miraflores Locks Panama Canal in position 08° 59.7N 79° 35.4W using International Navigation Rules, enroute San Diego. Sherman is DIW with both MDE's clutched in. Both MGT's are in standby, both SSDG's are paralleled providing electrical power. All standard navigation lights are energized and burning brightly, the Navigation and Anchor Detail is set. Material condition Yoke set throughout the ship with EMCON Delta in effect. Both boats are cradled for sea with SHE-2 as the ready boat. CGNR-6596 is hangered with primary, secondary, and heavy weather tie-downs in place. Sherman is under the OPCON, ADCON, and TACON of

COMPACAREA Alameda. 0000 Late Entry, GM2 absent without leave 0021 Exited Miraflores Locks 0028 Pilot boat alongside to starboard 0031 Disembarked 03 passengers to pilot boat 0032 Pilot boat away to starboard 0044 Passed beneath Thatcher Bridge 0047 Entered Pacific Ocean 0059 Pilot boat alongside to starboard 0100 Pilot disembarked to pilot boat 0102 Pilot boat away to starboard 0115 Secured the Navigation and Anchor Detail 0118 Singled up on NR 1 SSDG 0618 Observed sunrise, secured all standard navigation lights 1010 All small arms and ammunition present or accounted for 1200 Conducted test of ships alarms and whistle, test satisfactory 1220 Maneuvered to intercept contact in position 06° 44.7N 80° 49.7W 1249 Fire sighted on bow of M/V Ingram 1246 Fire on vessel reported out 1254 SHE-2 lowered to the main deck rail 1259 On scene in position 06° 44.7N 50° 49.6W 1308 The fire sighted was determined to be intentional to notify mariners of trouble 1225 SHE-2 lowered to the water with BM2 as coxswain, BM3, MK2, and MK3 as boat crew, and ENS, ENS, MK3, RD2, and SN as Boarding Team members 1327 SHE-2 away to port 1329 Boarding Team safely aboard M/V Ingrid 1107 SHE-2 alongside to port 1109 SHE-2 away to port after giving them one 12 volt battery 1447 SHE-2 alongside to port 1449 ENS, MK3, and RD2 safely aboard Sherman 1450 SHE-2 away to port 1507 SHE-2 alongside to port 1509 ENS safely aboard Sherman 1515 SHE-2 away to port 1532 SHE-2 alongside to port 1535 SHE-2 away to port with water 1537 SHE-2 alongside M/V Ingrid to deliver water 1540 SHE-2 alongside to port to receive a battery for M/V Ingrid 1542 SHE-2 away to port enroute M/V Ingrid 1556 SHE-2 alongside to port to receive 1 gallon of diesel fuel 1559 Traversed CGNR-6596 onto flight deck, primary and secondary tie downs in place

1609 SHE-2 away to port enroute M/V Ingrid to deliver fuel 1618 Set Flight Con 1 for launch, removed secondary tie downs from Helo 1620 Set the Helo Refueling Bill 1624 Secured the Helo Refueling Bill having transferred 27 gallons of JP-5 to Helo 1627 SHE-2 alongside to port, MK2 and SNBM safely aboard Sherman 1628 SHE-2 away to port 1629 Helo started engines and engaged rotors 1639 Removed primary tie downs from Helo 1640 Disengaged Talon, Helo is off the deck and away to port with 03 POB 1642 Secured Flight Con 1, set Flight Con 2 1656 SHE-2 alongside to port 1700 SHE-2 away to port 1812 Set Flight Con 1 for landing 1828 CGNR-6596 on deck 1829 Commenced Hot Refuel on deck with Helo, energized pump 1835 Secured pump after transferring 145 gallons of JP-5 to Helo 1838 Helo shut down engines and secured rotor 1839 Secured Flight Con 1 1845 Set the Towing Bill 1854 Placed both MDE's in Pilot House Control 1903 Passed the towline to M/V Ingrid 1907 Energized the lower towing light 1908 Commenced towing M/V Ingrid with 400 feet of towline at the stern rail, energized towing light. All towing lights are energized and burning brightly with the exception of the stern light 1920 Declutched NR2 MDE 1923 Placed NR1 MDE in Engine Room Control, SHE-2 alongside to port, disembarked MK1 1925 SHE-2 away to port 1926 SHE-2 alongside M/V Ingrid 1952 SHE-2 alongside to starboard, raised to main rail, energized stern towing 2021 SHE-2 cradled for sea 2208 Singled up on NR 1 SSDG

4 July - Underway in the Pacific Ocean in position 06° 55.7N 81° 33.2W using International Navigation Rules with M/V Ingrid in stern tow with 600ft of towline. BM3 and SNBM are on board M/V Ingrid Ships course 285° True, 287° Magnetic,

NR 1 MDE is online turning for 7 knots NR2 MDE and both MGT's are in standby, NR 2 SSDG is providing electrical power. All standard navigation lights are energized and burning brightly, Dog Zebra and material condition Yoke set throughout the ship with EMCON Delta in effect. Both boats are cradled for sea with SHE-2 as the ready boat. CGNR-6596 is hangered with primary, secondary, and heavy weather tie-downs in place. Sherman is under the OPCON, ADCON, and TACON of COMPACAREA Alameda. 0000 Late Entry, GM2 absent without leave 0200 Retarded clocks one hour to conform with +6 Sierra, time now 0100 0515 Paralleled both SSDG's 0520 Observed sunrise, secured all standard navigation and towing lights 0614 All small arms and ammunition present and accounted for 0800 SHE-2 lowered to the main deck rail 0807 SHE-2 lowered to the water with BM3 as coxswain, BM2 and MK3 as boat crew, MK1, BM3, and SNBM as passengers 0808 SHE-2 away to port enroute M/V Ingrid 0813 SHE-2 alongside M/V Ingrid, SNBM and BM3 safely aboard M/V, BM3 and SNBM safely aboard SHE-2 0815 SHE-2 enroute Sherman 0818 SHE-2 alongside to starboard, MK1, SNBM, and BM3 safely aboard Sherman 0819 SHE-2 away to starboard 0823 SHE-2 alongside to port 0824 SHE-2 raised to the main deck rail, coxswain and boat crew safely aboard Sherman 1210 SHE-2 lowered to water with BM3 as coxswain, SN, BM2, and MK3 as boat crew and SN and SN as passengers 1212 SHE-2 away to port enroute M/V Ingrid 1214 SHE-2 alongside M/V Ingrid 1215 SHE-2 alongside to starboard 1218 SHE-2 away to starboard 1300 Entered Panamanian Waters 1310 SHE-2 alongside to embark SN as passenger 1311 SHE-2 away to starboard 1340 SHE-2 alongside to starboard 1343 disembarked 03 POB 1415 SHE-2 alongside to port, raised to main deck rail

1640 Commenced preparations to break tow with M/V Ingrid 1646 SHE-2 away to port with BM1 as coxswain, BM3 and MK3 as boat crew 1649 Santa Maria on scene to assume tow, SHE-2 alongside M/V Ingrid 1652 Broke tow with M/V Ingrid 1656 Retrieved tow line from M/V Ingrid 1659 Santa Maria took M/V Ingrid in stern tow 1705 Secured the Towing Bill, SHE-2 enroute to Sherman with 02 Passengers 1710 SHE-2 alongside to port 1712 SHE-2 raised to the main deck rail, coxswain, boat crew, and 02 passengers safely aboard 1718 SHE-2 cradled for sea 1755 Observed sunset, energized all standard navigation lights 1920 Singled up on NR 2 SSDG

5 July - Underway in the Pacific Ocean in position 08° 28.3N 85° 31.8W using International Navigation Rules. Ships course 276° True, 276° Magnetic, NR 1 MGT is online turning for 22 knots NR2 MGT and both MDE's are in standby, NR 2 SSDG is providing electrical power. All standard navigation lights are energized and burning brightly, Dog Zebra and material condition Yoke set throughout the ship with EMCON Delta in effect. Both boats are cradled for sea with SHE-2 as the ready boat. CGNR-6596 is hangered with primary, secondary, and heavy weather tie-downs in place. Sherman is under the OPCON, ADCON, and TACON of COMPACAREA Alameda. 0000 Late Entry, GM2 absent without leave 0200 Clutched in NR2 MGT 0201 Declutched NR1 MGT, placed in standby 0540 Observed sunset, secured all standard navigation lights 0710 Clutched in both MDE's, declutched NR 2 MGT 0908 All small arms and small arms ammunition present and accounted for 0935 Commenced BECCE Drills 1023 Secured from BECCE Drills 1052 Paralleled both SSDG's 1200 Conducted test of the ships alarms and air whistle, test satisfactory 1315 Commenced

Collision Drill <u>1351</u> Secured from Collision Drill <u>1607</u> Traversed Helo onto flight deck with primary tie downs in place <u>1630</u> Set Flight Quarters Condition One for launch <u>1645</u> Helo engaged rotors and started engines <u>1653</u> Removed primary tie downs from Helo, Helo is off the deck and away to starboard <u>1655</u> Secured Flight Quarters Condition One, set Flight Condition Two <u>1810</u> Set Flight Quarters Condition One for launch <u>1815</u> Set Flight Condition Four for HIFR <u>1823</u> Helo is over the deck <u>1824</u> Passed refueling hose to Helo <u>1826</u> Recovered refueling hose from Helo <u>1827</u> Helo away to port <u>1828</u> Secured Flight Quarters Condition 4, set Flight Quarters Condition 1 <u>1829</u> Observed sunset, energized all standard navigation lights and restricted in ability to maneuver lights <u>1835</u> Helo commenced touch and go landings <u>1855</u> Helo is on deck after having completed 10 touch and go landings and 4 primary tie downs and Talon engagement <u>1858</u> Helo secured rotors and stopped engines <u>1900</u> Secured Flight Quarters Condition One, set the Helo Refueling Bill <u>1910</u> Secured Helo Refueling Bill after having transferred 165 gallons of JP-5 fuel to the Helo <u>1936</u> Singled up on NR 2 SSDG <u>2045</u> Traversed CGNR-6596 into the hanger, primary, secondary, and heavy weather tie downs in place

6 July - Underway in the North Pacific Ocean in position 11° 22.7N 91° 38.2W using International Navigation Rules. Ships course 296° True, 293° Magnetic, both MDE's are online turning for 16 knots both MGT's are in standby, NR 2 SSDG is providing electrical power. All standard navigation lights are energized and burning brightly, Dog Zebra and material condition Yoke set throughout the ship with EMCON Delta in effect. Both boats are cradled for sea with SHE-2 as the ready boat. CGNR-6596 is hangered with primary, secondary, and heavy weather tie-downs in place. Sherman is under the OPCON, ADCON, and TACON of COMPACAREA Alameda. GM2 is absent without leave 0600 Observed sunrise, secured all standard navigation lights 1005 All small arms and ammunition present or accounted for 1058 Received a registered EPIRB in position 12° 52N 95° 12W. Maneuvered to intercept. EPIRB is registered to S/V Further 1107 Paralleled both SSDG's 1108 Traversed CGNR-6596 onto flight deck, primary and secondary tie downs in place 1127 Set Flight Quarters Condition One for launch 1137 Helo engaged rotors and started engines 1144 Removed primary tie downs from Helo 1145 Helo off the deck and away to starboard with 04 POB 1148 Secured Flight Quarters Condition One, Set Flight Quarters Condition Two 1200 Conducted test of ships alarms and air whistle, test satisfactory 1206 CGNR-6596 and Navy P-3 on scene in position 12° 52N 95° 12W with negative sightings. Both aircraft commenced search pattern. 1225 Informed by CG PACAREA that second satellite pass detected no EPIRB signal 1317 Set Flight Quarters Condition One for landing 1323 Secured Flight Quarters Condition One, set Flight Quarters Condition Two. 1326 Received word from USCG PACAREA that location of EPIRB is in Colon, Panama. Sherman is released to continue transit. Navy P-3 is released 1337 Set Flight Quarters Condition One for landing 1347 Helo is on deck with Talon engaged with primary tie downs 1350 Commenced Hot Refuel on deck with Helo 1355 Secured Helo Refueling Bill after having transferred 170 gallons of JP-5 to the Helo 1358 Helo secured rotors and stopped engines 1359 Secured Flight Quarters Condition One 1732 Singled up on NR 2 SSDG

7 July - Underway in the North Pacific Ocean in position 14° 11.3N 97° 03.3W using International Navigation Rules. Ships course 297° True, 292° Magnetic, both MDE's are online turning for 16 knots both MGT's are in standby, NR 1 SSDG is providing electrical power. All standard navigation lights are energized and burning brightly, Dog Zebra and material condition Yoke set throughout the ship with EMCON Delta in effect. Both boats are cradled for sea with SHE-2 as the ready boat. CGNR-6596 is hangered with primary, secondary, and heavy weather tie-downs in place. Sherman is under the OPCON, ADCON, and TACON of COMPACAREA Alameda. GM2 is absent without leave 0200 Retarded clocks 1 hour to conform with +7 Tango 0510 Observed sunrise, secured all standard navigation lights 0741 Clutched in both MGT''s, declutched both MDE's 0837 Paralleled both SSDG's 0902 All small arms and ammunition present or accounted for 1200 Conducted test of ships alarms and air whistle, test satisfactory 1205 Clutched in both MDE's 1207 Declutched both MGT's 1237 Singled up on NR 1 SSDG 1352 SHE-2 lowered to the main deck rail 1405 SHE-2 away to port with BM3 as coxswain BM2, SNBM, MK2 and GM3 as boat crew 1409 Declutched both MDE's 1443 Clutched in both MDE's 1447 SHE-2 alongside to port, transferred 1 POB 1449 SHE-2 raised to main deck rail, boat crew safely aboard 1825 Observed sunset, energized all standard navigation lights

8 July - Underway in the North Pacific Ocean in position 17° 34.1N 103° 38.2W using International Navigation Rules enroute San Diego, CA. Ships course 298° True, 292° Magnetic, both MDE's are online turning for 16 knots both MGT's are in standby, NR 1 SSDG is providing electrical power. All standard navigation lights are energized and burning brightly, Dog Zebra and material condition Yoke set throughout the ship with EMCON Delta in effect. Both boats are cradled for sea with SHE-2 as the ready boat. CGNR-6596 is hangered with primary, secondary, and heavy weather tie-downs in place. Sherman is under the OPCON, ADCON, and TACON of COMPACAREA Alameda. GM2 is absent without leave 0529 Observed sunrise, secured all standard navigational lighting 0928 Operating in various plant configurations due to unrestricted BECCE Drills 0945 All small arms and ammunition present or accounted for 1018 Secured from unrestricted BECCE Drills, both MDE's are clutched in 1535 Paralleled both SSDG's 1854 Observed sunset, energized all standard navigation lights 1959 Singled up on NR 1 SSDG

9 July - Underway in the North Pacific Ocean in position 21° 00.2N 108° 46.2W using International Navigation Rules enroute San Diego, CA. Ships course 303° True, 294° Magnetic, both MDE's are online turning for 16.5 knots both MGT's are in standby, NR 1 SSDG is providing electrical power. All standard navigation lights are energized and burning brightly, Dog Zebra and material condition Yoke set throughout the ship with EMCON Delta in effect. Both boats are cradled for sea with SHE-2 as the ready boat. CGNR-6596 is hangered with primary, secondary, and heavy weather tie-downs in place. Sherman is under the OPCON, ADCON, and TACON of COMPACAREA Alameda. GM2 is absent without leave 0547 Observed sunrise, secured all standard navigation lights 0645 Maneuvered to intercept USS Scout for passenger transfer in position 21° 03.7N 110° 13.0W 0659 On scene with USS Scout (MCM-8) in position 22° 02.3N 108° 13.3W 0703 Paralleled

SSDG's 0705 SHE-2 lowered to the main deck rail, coxswain and boat crew safely aboard SHE-2 0709 SHE-2 lowered to the water and away to port with BM as coxswain, BM3, and MK3 as boat crew enroute USS Scout (MCM-08) 0712 SHE-2 alongside USS Scout (MCM-8), 3 passengers safely on board SHE-2 0714 SHE-2 away enroute Sherman with 3 passengers 0716 SHE-2 alongside to port 0717 SHE-2 disembarked with 3 POB safely on board Sherman 0733 SHE-2 alongside t port 0737 02 passengers disembarked to SHE-2, SHE-2 away to port 0741 SHE-2 alongside USS Scout (MCM-8) 0743 Disembarked 02 passengers safely aboard USS Scout (MCM-8), SHE-2 away to port 0746 SHE-2 alongside to port 0748 SHE-2 raised to the main deck rail with coxswain and boat crew safely aboard 0756 SHE-2 is ready and cradled for sea 0722 Late Entry – MS2, USN reported from USS Scout (MCM-8) for medivac 0815 Set Flight Quarters Condition One for launch 0834 Helo engaged rotors and started engines 0800 Late Entry – Traversed Helo onto flight deck 0840 Helo is off deck and away to port with 04 POB 0845 Secured Flight Quarters Condition One, set Flight Quarters Condition Two 1001 Set Flight Quarters Condition One for landing 1020 Helo is on deck with Talon engaged 1023 Helo disengaged rotors and stopped engines 1024 Helo Crash Drill 1035 Secured from Helo Crash Drill and Flight Quarters Condition One, set the Helo Refueling Bill 1041 Secured Helo Refueling Bill after having transferred 125 gallons of JP-5 to Helo 1200 Conducted test of ships alarms and whistle, test satisfactory 1417 Singled up on NR 1 SSDG 1633 Late Entry – CGNR-6596 traversed into thee hanger, primary, secondary, and heavy weather tie downs in place 1920 Secured from Commanding Officers Non-Judicial Proceedings. All charges were dismissed 1921 Observed sunset, energized all standard navigation lights

10 July - Underway in the North Pacific Ocean in position 25° 01.8N 113° 29.3W using International Navigation Rules enroute San Diego, CA. Ships course 320° True, 310° Magnetic, both MDE's are online turning for 16 knots both MGT's are in standby, NR 1 SSDG is providing electrical power. All standard navigation lights are energized and burning brightly, Dog Zebra and material condition Yoke set throughout the ship with EMCON Delta in effect. Both boats are cradled for sea with SHE-2 as the ready boat. CGNR-6596 is hangered with primary, secondary, and heavy weather tie-downs in place. Sherman is under the OPCON, ADCON, and TACON of COMPACAREA Alameda. GM2 is absent without leave 0551 Observed sunrise, secured all standard navigational lighting 0830 All small arms and ammunition present or accounted for 1200 Conducted test of ships alarms and air whistle, test satisfactory 1947 Observed sunset, energized all standard navigational lighting 2126 Commenced preparations for entering port in San Diego, California

11 July - Underway in the North Pacific Ocean in position 30° 23.2N 116° 43.0W using International Navigation Rules enroute San Diego, CA. Ships course 344° True, 333° Magnetic, both MDE's are online turning for 16 knots both MGT's are in standby, NR 1 SSDG is providing electrical power. All standard navigation lights are energized and burning brightly, Dog Zebra and material condition Yoke set throughout the ship with EMCON Delta in effect. Both boats are cradled for sea with SHE-2 as the ready boat. CGNR-6596 is hangered with primary, secondary, and heavy weather tie-downs in place.

Sherman is under the OPCON, ADCON, and TACON of COMPACAREA Alameda. GM2 is absent without leave 0555 Observed sunrise, secured all standard navigation lights 0642 Clutched in NR1 MGT, declutched both MDE's placed in standby. Paralleled both SSDG's 0700 Conducted time tick, conducted test of ships alarms and whistle, test satisfactory 0740 Set Flight Quarters Condition One for launch 0747 Clutched in both MDE's 0816 Helo engaged rotors and started engines 0819 Helo is off deck and away to port with 03 POB 0821 Secured from Flight Quarters Condition One 0904 Captain Ryan assumed the Conn QMC Eagleton assumed the deck SNQM assumed the QMOW 0918 All small arms and small arms ammunition present and accounted for 0945 SHE-2 lowered to the water with BM3 as coxswain, SN, SA, SN as boat crew 0951 Set Special Sea Detail 0954 SHE-2 away to port 0958 Lowered and tested Bow Prop, test satisfactory, placed in Pilot House Control 0954 Secured Special Sea Detail, Moored FISC Pier, San Diego, CA. 1245 Set the Special Sea Detail 1255 Singled up on all mooring lines 1335 Lowered and tested Bow Prop, test satisfactory, placed in Pilot House Control 1336 NR 2 MDE Clutched in 1337 NR 1 Clutched in, both MDE's placed in Pilot House Control 1342 Underway in San Diego Harbor, Mr. (name redacted) assumed the Conn 1348 Secured from Special Sea Detail with the exception of Nav and Anchor Detail 1404 Commenced approach on anchorage 1417 Anchored in anchorage in Sa Diego Harbor in position (redacted) with 1 shots of chain on deck to starboard anchor to a sand bottom in 22 feet of water using the following bearings, 142° True to L, 107° True to Z and 42.5 yards to Z. Hoisted dayshapes 1420 Commenced heaving around on starboard anchor 1441 Anchors Aweigh, underway in San Diego harbor enroute Alameda, CA.

Secured all dayshapes 1433 SHE-2 alongside to port 1436 SHE-2 raised to the main deck rail. Boat crew safely aboard Sherman 1439 SHE-2 cradled 1500 Crossed Line of Demarcation outbound to sea. Using International Rules 1509 Secured Navigation and Anchor Detail 1518 Clutched in both MGT's, declutched both MDE's 1636 Clutched in both MDE's, declutched both MGT's 1723 Set the UNREP Bill 1752 Commenced approach on the USNS Guadalupe (TAO-200) 1755 Alongside the USNS Guadalupe for refueling 1758 USNS Guadalupe passed messenger 1801 Passed bridge to bridge phone line 1802 USNS Guadalupe passed spanwire 1805 Tensioned spanwire 1807 Received coupled fuel hose 1814 Commenced fueling 1939 Completed fueling having received 107,917 gallons of JP-5 fuel 1940 Uncoupled fueling hose and passed to USNS Guadalupe 1941 Detensioned spanwire 1943 Recovered bridge to bridge phone line, broke away from USNS Guadalupe 1947 Secured from UNREP Bill 1955 Singled up on NR 2 SSDG 2004 Observed sunset, energized all standard navigation lights

12 July - Underway in the North Pacific Ocean in position 32° 42.4N 119° 40.5W using International Navigation Rules enroute Alameda, CA. Ships course 283° True, 270° Magnetic, NR2 MDE is online turning for 12 knots, NR 1 MDE and both MGT's are in standby, NR 2 SSDG is providing electrical power. All standard navigation lights are energized and burning brightly, Dog Zebra and material condition Yoke set throughout the ship with EMCON Delta in effect. Both boats are cradled for sea with SHE-2 as the ready boat. Sherman is under the OPCON, ADCON, and TACON of COMPACAREA Alameda. GM2 is absent without leave 0605 Observed sunrise, secured all standard navigation lights 0631 Clutched in NR 1 MDE 0755 Paralleled both SSDG's 0830 All small arms and ammunition present or accounted for 0900 Commenced MK75 76mm and CWIS Gunnery Exercise 0917 Commenced Fire 0928 Secured MK75 76mm having expanded 36 rounds with no apparent casualties 0956 Secured from Gunnery Exercise after having expanded 300 rounds of CWIS with no apparent casualties 1200 Conducted test of ships alarms and air whistle, test satisfactory 1314 Commenced GE Drill 1429 Secured from GE Drill 1500 Singled on NR 2 SSDG 2038 Observed sunset, energized all standard navigation lights 2141 Declutched the NR 1 MDE, placed in standby 0800 Late Entry – Commenced preparations for entering port in Alameda, CA.

13 July - Underway in the North Pacific Ocean in position 36° 34.2N 122° 07.4W using International Navigation Rules enroute Alameda, CA. Ships course 313° True, 299° Magnetic, NR2 MDE is online turning for 12.5 knots, NR 1 MDE and both MGT's are in standby, NR 2 SSDG is providing electrical power. All standard navigation lights are energized and burning brightly, Dog Zebra and material condition Yoke set throughout the ship with EMCON Delta in effect. Both boats are cradled for sea with SHE-2 as the ready boat. Sherman is under the OPCON, ADCON, and TACON of COMPACAREA Alameda. GM2 is absent without leave 0247 Visibility decreased to under 2 miles energized sound signal for underway making way 0302 Visibility increased, secured sound signal 0505 Clutched in NR 1 MDE 0530 Visibility decreased to under 2 miles energized sound signal for underway making way 0550 Observed sunrise 0740 Set the Navigation and Anchor Detail 0801 Visibility increased to 84,000 yards, secured fog signals 0810 Declutched

NR1 MDE 0811 Coast Guard P/B alongside to starboard 0814 All small arms and small arms ammunition present and accounted for 0815 Embarked 06 passengers 0818 Clutched in NR1 MDE 0805 Late Entry – Crossed Line of Demarcation for San Francisco Bay, shifted to Inland Navigational Rules 0830 Passed beneath Golden Gate Bridge, entered San Francisco Bay 0842 Secured all standard navigational lights 0857 Passed beneath Oakland Bay Bridge, entered Oakland Estuary 0912 Placed both MDE's in Pilot House Control 0924 Set Special Sea Detail 0941 Lowered and tested Bow Prop, test satisfactory, placed in Pilot House Control 0950 Moored port side Coast Guard Island Alameda, California 0954 Secured Special Sea Detail 0957 Raised and secured Bow Prop 0957 Declutched both MDE's 1011 OOD shifted the watch from the Pilot House to the quarterdeck 1015 Liberty Granted to all hands with the exception of duty section one, to expire on board no later than 0645 17 July for all hands 1016 GM2 returned onboard 1025 (rate and name redacted) reported PCS from Activities New York, SNQM reported PCS from RESTRACEN Yorktown, VA. (rate and name redacted) reported PCS from TRACEN Cape May, NJ, SK1 reported PCS from Alameda, CA. 1044 GM1 reported PCS from NESU Alameda 1111 TC1 reported PCS from CG AIRSTA Travis City 1120 SK2 departed PCS to Group Mobile 1122 SK3 departed PCS to ISC Seattle, WA. 1136 Shifted from ships power to shore power 1218 CWO reported PCS from CG Academy, New London, CT. 1620 BMC reported PCS from ANT Jacksonville Beach, FL. 1832 Observed sunset, energized all deck, anchor, and aircraft warning lights 2247 SA arrived PCS from TRACEN Petaluma

SOURCES AND REFERENCES

Around the World: 25 years of service as an officer and enlisted man in the U.S. Army and U.S. Coast Guard
  By Edward L. Semler Jr.
  ISBN: 978-0-615-77701-6

Associated Press article "Ship Sinks In Persian Gulf Spilling Smuggled Iraqi Oil"
  19 June 2001

Discussions with Dave Ryan, John Young, Ben Strickland, Andrew Vandewarker, and Dave Socci.

Kuwaiti divers made honorary Desert Cats
  By Catherine Desert

Navy Times article "Coast Guard Uniquely Suited For Mission"
  16 April 2001
  By William H. McMichael

Navy Times article "Cutters Circle The Globe For Exercise"
  13 August 2001
  By Patricia Kime

Photos courtesy of Edward Semler, John Young, Noah Alberici, Chris Randolph, Andrew Vandewarker, Ben Strickland, Chris Guinther, Joseph Prince, and Kort Huettinger

Operational Summary and Lessons Learned For PACMEF 01-1 Deployment of CGC Sherman
24 August 2001

The Cargo Letter at www.cargolaw.com article "Bumper Cars"
By The law offices of Countryman & McDaniel

The Sentinel Staff article "Embassy Survivor Has Ties To Orange"
12 August 1998
By Pedro Ruz Gutierrez

USCG Message Z121543Z JUL01 from COMPACAREA COGARD Alameda

USCGC Sherman newsletters January through August 2001

USCGC Sherman ships logs 13 January to 13 July 2001 (redacted)

USCG Unit Commendation Award dated 3 July 2001 to USCGC Sherman

USN Message R241301Z JUL01 from COMUSNAVCENT

USS Alexandria (SSN-757) 2001 Command History
29 April 2002

USS Ardent (MCM-12) 2001 Command History
30 January 2002

USS Arleigh Burke (DDG-51) 2001 Command History
05 March 2002

USS Carr (FFG-52) 2001 Command History
01 March 2002

USS Enterprise (CVN-65) 2001 Command History
26 August 2002

USS Harry S. Truman (CVN-75) 2001 Command History
21 March 2002

USS Mitscher (DDG-57) 2001 Command History
27 April 2002

USS Stethem (DDG-63) 2001 Command History
18 January 2002

USS Stump (DD-978) 2001 Command History
01 March 2002

ABOUT THE AUTHOR

Ed Semler retired from the United States Coast Guard in December of 2007 with over 25 years of military service in both the United States Army and United States Coast Guard. In the United States Army he was an enlisted man and was honorably discharged as a specialist four (E-4). While in the United States Coast Guard he was enlisted, obtaining the rank of master chief petty officer (E-9), was commissioned as an officer, and retired as a lieutenant (O-3E).

He served from 20 October 1999 to 01 August 2001 aboard *Sherman* as the engineering division main propulsion senior chief petty officer. As a collateral duty he was also the command enlisted advisor (CEA). He promoted to master chief machinery technician on 01 March 2001 aboard *Sherman* while she was underway on her historic circumnavigation deployment.

His other publication is "Around The World: 25 Years Of Service As An Officer And Enlisted Man In The U.S. Army And U.S. Coast Guard," which is a memoir of his military service.

He resides in Schulenburg, Texas with his wife Jana, a retired Air Force senior master sergeant. Feel free to contact him at mkcm378@gmail.com or www.edsemler.com

www.ingramcontent.com/pod-product-compliance
Lightning Source LLC
Chambersburg PA
CBHW060149050426
42446CB00013B/2742